普通高等教育“十三五”规划教材暨智能制造领域人才培养规划教材

机械设计制造及其自动化专业英语

Professional English of Mechanical Design, Manufacturing and Automation

主　编　胡珊珊
副主编　王　艳
参　编　胡映宁　王小纯　陈远玲
　　　　汤宏群　孙屹博　郑冬锐

华中科技大学出版社
中国·武汉

内容简介

本教材的阅读材料主要选材于经典的国外原文教材、国外学术著作和期刊；为了增加学习趣味性，增加了最新的网络信息和流行的机械类专业杂志的内容。内容的选择既考虑了专业知识的常见性，又兼顾专业英语教学过程与基础英语的过渡，选用语言地道、形式多样但不涉及过于繁杂和深奥理论的文献。通过学习本教材，读者可掌握通过多种渠道（包括专业学术期刊，硕士、博士论文，机械类全英文报刊，机械产品广告，通过网络视频和文字发布的最新科研成果，与机械设计制造相关的新闻报道等等）扩展专业英语知识的方法，十分适合在现今发达的网络和数字化信息平台下获取新的专业英语知识。

本教材分为六章，每一章围绕一个主题展开，第一章介绍本专业，第二章介绍机械设计制造史，第三章介绍如何阅读及理解专业文献，第四章介绍随时随地学习专业英语，第五章介绍实验中的英语，第六章介绍以多种方式使用专业英语。其中第三章可以从学习阅读及理解专业文献的角度，以章节形式进行；也可以从传统的机械设计制造领域分类的角度，以 TOPIC 形式进行。这将给使用本教材的教师提供多种可选择的教学方案，并通过课后练习的方式，引导教师、学生及使用本教材自学的读者学习如何阅读及理解英文专业文献，形成并发展自己的方法，可以达到“授人以渔”的效果。

图书在版编目(CIP)数据

机械设计制造及其自动化专业英语/胡珊珊主编．—武汉：华中科技大学出版社，2018.3
(2025.2重印)
普通高等教育“十三五”规划教材暨智能制造领域人才培养规划教材
ISBN 978-7-5680-3688-7

Ⅰ.①机… Ⅱ.①胡… Ⅲ.①机械制造-英语-高等学校-教材 ②自动化技术-英语-高等学校-教材 Ⅳ.①TH16 ②TP

中国版本图书馆 CIP 数据核字(2018)第 041093 号

机械设计制造及其自动化专业英语 胡珊珊 主编
Jixie Sheji Zhizao jiqi Zidonghua Zhuanye Yingyu

策划编辑：万亚军 封面设计：原色设计
责任编辑：程 青 责任校对：李 琴
责任监印：周治超
出版发行：华中科技大学出版社(中国·武汉) 电话：(027)81321913
武汉市东湖新技术开发区华工科技园 邮编：430223
录 排：华中科技大学惠友文印中心
印 刷：广东虎彩云印刷有限公司
开 本：710mm×1000mm 1/16
印 张：10.25
字 数：216 千字
版 次：2025 年 2 月第 1 版第 6 次印刷
定 价：35.00 元

前　言

随着国际化的深入和日益广泛的国内外交流与合作，我国制造业迅猛发展，政府部门对制造业越发重视，制造业人才的需求量逐年增加，急需具备良好的国际沟通能力和紧跟技术发展动向、技术敏锐力高的高级技术通才。这意味着该类人才不仅要熟悉相关领域的专业技术，并能凭借其良好的国际沟通能力，掌握前沿技术并将其应用到企业发展中。

编者在多年的机械设计制造双语专业课、机械设计制造及其自动化专业英语的教学过程中发现，在从基础英语到专业英语的教学过程中缺乏有效过渡，导致很多学生认为专业英语既艰涩又枯燥。因此，从教学实践的角度出发，编写一本内容全面、可读性强、适合课堂教学的机械设计制造技术方面的专业英语教材十分必要。

在编撰本教材的过程中，编者有幸在美国密歇根大学访学一年。在访学过程中发现，在日常生活、课程学习和学术交流中有很多需要掌握的专业英语的表达方法。编者将这些看似简单却必要的表达方法放入本教材中，以期引起从业教师及教学教育专家的注意，达到抛砖引玉的效果。

感谢密歇根大学 S. M. Wu Manufacturing Research Center 主任倪军教授为本教材的编撰提供了优良的研究环境及丰富的学术资源。在本书编写过程中，Katherine Sholder 对全文的语法和表达进行了仔细的修订，兰州理工大学郑玉巧、吉林大学冀世军、华侨大学尤芳怡、南京航空航天大学朱栋、吉林大学张秀芝等老师对编写结构和内容给予了启发和帮助，同时赵国林、阮帆、熊飞翔等人也给予了协助，在此一并感谢。

限于时间和学术水平，本书难免存在错误及不足之处，恳请广大读者批评指正。

编　者

2017 年 9 月

目　录

Chapter 1 Know about Your Major

1.1 An Overview of Majors/Programs in Universities All Over the World

Before we start to study this book, it is necessary to have an overview of similar majors/programs in universities all over the world. This will give you a basic understanding of why students choose this major, what students learn, and what they can do after graduation.

University

University of Limerick (UL) (Ireland)

Major

Bachelor/Master of Engineering in Mechanical Engineering

About You

This is an ideal programme for you if you are interested in solving problems using mathematics and science. If you think you might enjoy exploring areas such as mechanical design, energy systems and materials, then Mechanical Engineering at UL might be a good choice for you.

Why Study Mechanical Engineering at UL?

Mechanical Engineering at UL now offers an integrated bachelor/master of engineering programme. The entry route to both is through LM116 but in year 3 students have the choice to decide the bachelor or master of engineering programme.

- Bachelor of engineering in mechanical engineering (4 years in duration)
- Master of engineering in mechanical engineering (5 years in duration)

Mechanical engineering is a very broad-based discipline and students following the degree programme are prepared for careers in many industrial sectors, including such diverse areas as energy, chemical processing, research, manufacturing, design consultancy, material processing and aviation. The mechanical engineering degree programme aims not only to give you a thorough background in fundamental mechanical engineering subjects but also allows specialisation in one of a number of areas of particular relevance to Irish industry.

Mechanical Engineering at UL adheres to traditional guidelines set down by the professional engineering institutions (such as Engineers Ireland and IMechE) and requires you to have an aptitude for mathematics and problem solving.

Mechanical Engineering at UL is an honours degree programme accredited by Engineers Ireland (www. engineersireland. ie), and the qualifications of graduates are recognised world-wide through international accords.

Mechanical Engineering at UL now offers students the opportunity to undertake an integrated Master of Engineering programme. Student can choose this route in year 3 and will study for a further 2 years graduating with an M. E. in Mechanical Engineering. For more information, please search www. scieng. ul. ie/ schoolsdepartments/.

Course Structure

Common Entry to Biomedical, Civil, Design & Manufacture and Mechanical Engineering

The bachelor of engineering programme is of four years in duration and is divided into two parts.

Part Ⅰ

Part Ⅰ, which comprises the first year of study, provides you with a foundation in the fundamental engineering subjects and makes up for variations in the background of individual student: mathematics, computing, engineering mechanics, physical chemistry, electrical principles, fluid mechanics, production technology, the engineering profession.

Part Ⅱ

Part Ⅱ comprises years 2, 3 and 4 and you will generally study five modules per semester. You will study all the fundamental subjects of mechanical engineering—mathematics, mechanics of solids, design, mechanics of fluids, thermodynamics, dynamics of machines and control.

At the end of year 2, you are placed in industry for an eight-month cooperative education period. This period provides experience of the practice and application of mechanical engineering in an industrial environment. You will then return to the university for the latter half of the third year and start to specialise.

This programme offers a broad-based course in mechanical engineering. In addition, in the final year, you can specialise in thermofluids, mechanics of solids or energy by choosing appropriate final year electives.

An important aspect of this programme is the final year project completed in year 4. This is an individual project assigned to you at the end of year 3 giving you almost 12 months to undertake. The project is a major piece of work and involves the preparation of a report detailing all aspects of the project. It will provide you with the opportunity to demonstrate your ability to work as a professional engineer and to incorporate the knowledge you have gained over the previous three years. Many students are proud to show the work at subsequent job interviews.

For more information, please search www. scieng. ul. ie/ schoolsdepartments/.

Entry Requirements

Applicants are required to hold at the time of enrolment the established leaving certificate (or an approved equivalent) with a minimum of six subjects which must include: two H5 grades and four O6 grades or four H7 grades (H means higher level and O mean ordinary level). Subjects must include mathematics, Irish or another language and English.

In addition, applicants must hold a minimum grade H4 in mathematics and grade O6/H7 in one of the following: physics, chemistry, physics with chemistry, engineering, technology, design & communication graphics/ technical drawing, biology, agricultural science, applied maths, construction studies.

A special mathematics (higher level) examination will be offered at UL following the leaving certificate results for those students who did not achieve the mathematics requirements. We welcome applications from mature students. Mature applicants must apply through the Central Applications Office (CAO) by 1 February.

Career Prospects

Recent graduates of the programme have found jobs in the following areas.

- Automotive and manufacturing engineering
- Offshore① engineering
- Aeronautical② engineering
- Pharmaceutical③ and biomedical④ industries

① offshore [ˌɒfˈʃɔː] *adj.* 离岸的；近海的

② aeronautical [ˌeərəˈnɔːtɪkl] *adj.* 航空的；航空学的

③ pharmaceutical [ˌfɑːməˈsuːtɪkl] *adj.* 制药(学)的

④ biomedical [ˌbaɪəʊˈmedɪkl] *adj.* 生物医学的

- Optimisation and design of energy systems
- Materials and structural analysis
- Engineering consultancy
- Project management
- Control of chemical and pharmaceutical
- Bioengineering① and life sciences
- Research and development

University

University of Michigan (USA)

Major

Mechanical Engineering

Undergraduate Programs & Degrees

First year undergraduate engineering students who have not transferred from another college or university will enter the College of Engineering without declaring a specific engineering major. Students do not typically declare a major until sophomore year. College of Engineering offers 17 undergraduate programs of study that lead to a Bachelor of Science degree.

Undergraduate Program (Mechanical Engineering) Overview

Mechanical Engineering at Michigan is an undergraduate program on the move. Since 2002, They've had a boom of new faculty members and research funding, while maintaining their consistent leadership as one of U. S. News and World Report's top five ranked ME (mechanical engineering) programs in the country. Their multidisciplinary② approach to research and learning strikes a unique balance of trend-setting③ research and challenging coursework that is highly respected around the world. At Michigan, ME is assembling the finest young undergraduate talent in the country.

What's Involved in an ME Degree at the University of Michigan?

At the University of Michigan, the Bachelor of Science in Engineering degree

① bioengineering [ˌbaɪəʊˌendʒnɪərɪŋ] *n*. 生物工程;[生物物理] 生物工程学
② multidisciplinary [mʌltidɪsəˈplɪnəri] *adj*. 有关各种学问的,包括多种学科的
③ trend-setting 创造潮流的,引领潮流的

(BSE) in ME provides students with an excellent foundation in the core technical competencies of the following disciplines: thermal and fluid sciences, solid mechanics and materials, dynamics and control. Within each of these disciplines, students will complete rigorous coursework that follows the traditional classroom format of lectures, discussions, homework, projects and exams. In addition, an array of technical electives are offered to enable students to tailor their mechanical engineering education to best suit their career goals.

During the sophomore and junior year, ME students will participate in design courses where they will work together as a team and apply the knowledge learned in their core subjects to develop a product, design and model it, and physically build it in machine shop. During their senior year, ME students will enroll in a senior capstone design course in which each team is required to complete a semester long design project developed for them by a company or researching faculty. The laboratory sequences are completed during the students' junior and senior years and involve working on a team to conduct industry-related experiments and perform data analysis. During both the design/manufacturing and laboratory courses, students are further prepared for successful careers and leadership positions by building extensive teamwork, report writings and presentation skills.

In addition to the regular BSE degree in mechanical engineering, there are numerous other programs offered to enrich education, such as dual-degrees (ME degree and a second degree from another engineering program), sequential undergraduate/graduate studies (SUGS), the engineering global leadership program (EGL), study abroad (listed on CoE minors), and independent study opportunities with ME faculty.

Students who do well in their undergraduate programs are encouraged to consider graduate work and may take some of their electives in preparation for graduate study. For more information please refer to the Graduate Handbook.

University

Massachusetts Institute of Technology

Major

Mechanical Engineering

The Educational Objectives of the Program Leading to the Degree Bachelor of Science in Mechanical Engineering

Within a few years of graduation, a majority of graduates will have completed or be progressing through top graduate programs; advancing in leadership tracks in industries, non-profit organizations, or the public sectors; or pursuing entrepreneurial ventures. In these roles they will: ① apply a wide working knowledge or technical fundamentals in areas related to mechanical, electromechanical, and thermal systems to address the needs of customers and society; ② develop innovative technologies and find solutions to engineering problems; ③ communicate effectively as members of multidisciplinary teams; ④ be sensitive to professional and societal contexts and committed to ethical actions; ⑤ lead in the conception, design, and implementation of new products, processes, service and systems.

Students are urged to contact the MechE Undergraduate Office as soon as they have decided to enter mechanical engineering so that faculty advisors may be assigned. Students, together with their faculty advisors, plan a program that best utilizes the departmental electives and the 48 units of unrestricted electives available in the Course 2 degree program. This program is accredited by the Engineering Accreditation Commission of ABET as a mechanical engineering degree.

University

Princeton University

Programs

Mechanical and aerospace engineering

Mechanical and aerospace engineers design, build and test devices and vehicles, such as cars, aircraft, satellites, engines, robots and control systems. Increasingly, electronics, computers, and mechanical devices are more and more integrated, mechanical and aerospace engineers must have a very wide knowledge and training in order to perform their jobs at the highest level. This university's program emphasis is to provide an education in the fundamentals of engineering which require the understanding and application of physical phenomena. They follow a broad system approach, where engineering decisions are made with a full appreciation of the opportunities and limitations presented by advanced technologies and their integration.

The department recognizes that students follow a variety of career paths. Some enter industries directly as practicing engineers while others continue their studies in graduate school in the fields of engineering or applied science. Others follow programs in preparation for careers in business, law or medicine. Some students enter the military. For the class of 2015, 17% of graduates decided to continue their studies in graduate school in engineering such as Cal Tech, Stanford, Yale, and Cornell. 60% chose a technical career in industries such as Ford Motor Company, Creative Edge Products, Dassault Falcon Jet, Lockheed Martin, Virgin Galactic, Facebook, Delta Airlines, SpaceX, Universal Creative, Ad Energy, Pratt and Miller, JRI-America and Telnyx. Some entered a career in the military with the U. S. Navy and the Naval Surface Warfare Center. Others began careers in management consulting, finance or environmental consulting with such companies as Boston Consulting Group, Strategy & Company, Black Rock, UBS, Oliver Wyman, Bridgewater Associates and McKinsey.

They respond to their students' varied interests by offering interdepartmental programs and topical programs. Sufficient flexibility is provided to meet a range of career objectives while providing a foundation of the engineering disciplines and associated problem solving skills.

1.2 Extended Exercise—Know about Your Knowledge of Mechanical Engineering

Please draw a mechanical engineering knowledge structure by yourself. Compare this to the previous universities' requirements and goals, and find out how excellent you are!

An example is shown in Figure 1-1.

1.3 Extended Exercise—Recognize Yourself

(1) It is generally acknowledged that most people represent information in their minds in one of three ways: as semantic information (words), as graphical information (visual images), and as analytical information (equations or relationship). Choose one of your friends in the classroom, write down the way you think he/she is good at presenting the information in Table 1-1, and write down the way you think you are good at presenting the information. What's the difference?

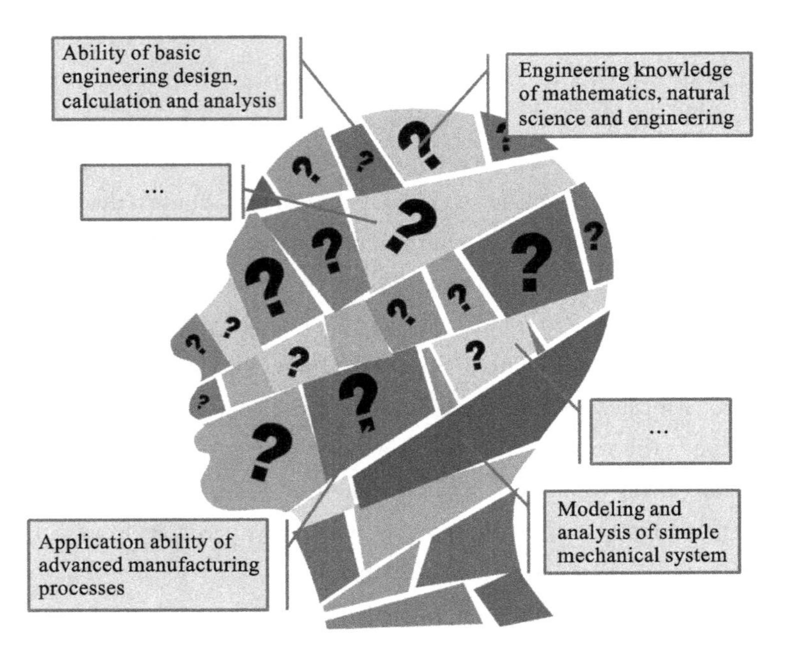

Figure 1-1　An Example of Mechanical Engineering Knowledge Structure

Table 1-1　How do You and Your Friend Present Information in Minds?

I think my friend ________ is good at ________.	I think I am good at ________.
1.	1.
2.	2.
3.	3.

(2) Define yourself by the poster!

There are some posters (Figure 1-2) describing engineers. Draw your own poster and describe yourself, then show it to everyone in the class!

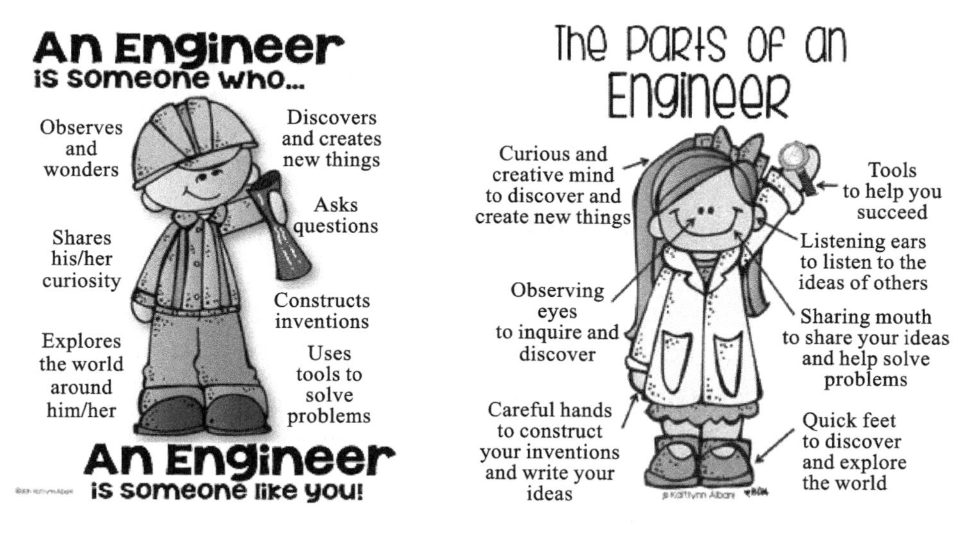

Figure 1-2 Posters Describing Engineers

Chapter 2 Know about the History of Mechanical Design and Manufacturing

2.1 Fire is Stronger than Blood and Water—Steam Power

I sell here what all men desire—Power!

Matthew Boulton (1728-1809)

Industrialist, manufacturer of steam engines

More than any other invention, more than any other device, the industrial revolution is known for the steam engine. While all of the previous developments like the manufactories or the first factories are relevant to the industrial revolutions, probably the most significant change is the development of the steam engine. The ancient Greeks were the first to mention steam-powered movements, but these were gadgets with no practical value. Only during the late seventeenth century did steam power become available. This availability of unprecedented power at almost any location changed manufacturing and the world more and faster than any other preceding developments in the history of mankind did.

Thomas Savery developed the first steam engine—or more correctly, steam pump—and patented it in 1698. The technology was still in its infant stage. There were no pistons but merely a set of valves to let steam in the chamber and water in and out of it. Hot steam was filled into a chamber and then cooled down inside it. The condensing steam created a vacuum, sucking up water through an attached pipe. This sucked-up water was then pushed upward with the inflowing steam of the next cycle. Since this engine did not create any mechanical movements but merely pumped water, it was not useable for manufacturing purposes. Additionally, due to the constant heating and cooling of the cylinder and the steam, lots of energy was wasted, and the engine worked very inefficiently. The only use of the engine was to pump up water, for which it was used with limited success to drain water out of mines. Hence, it was also advertised as a *miner's friend*. However, there were numerous problems with the engine. The engine was very inefficient, requiring expensive fuel and in need of constant maintenance. The engine was also only able to pump water from 9 meters below to about 10 meters above the engine. Hence, in order to drain[①] a mine, the engine had to be installed inside of the mine, creating lots of additional problems like providing fresh air and ventilation[②] of the exhaust. Finally, the engine also had a tendency to explode due to weak joints, and therefore, the miners might have been less excited about their new *miner's friend* deep down in the mine shafts. Overall, only 3% of all steam engines installed in Great Britain during the eighteenth century were Savery steam pumps.

A significant step forward was made by Thomas Newcomen, who, in 1712, built the first actual steam engine that converted steam into mechanical movement. The condensing steam no longer sucked up water but, rather, was sucked in a piston. While the constant heating and cooling of the cylinder still made the engine very inefficient, it was now possible to install the engine on the surface and still bring the movement through shafts and chains to the pumps inside of the mine. The engine was able to raise water up more than 30 meters. The engine worked with less steam pressure than the Savery engine and hence was unlikely to explode, surely a comforting fact to the operators of the engine. Subsequently, the Newcomen steam engines were very successful, and more than two-thirds of all engines installed in Great Britain during the eighteenth century were Newcomen steam engines. Hence, Newcomen steam engines were the first successful and widely used steam engine.

A model of the Newcomen steam engine was brought for repair to the instrument maker James Watt (Figure 2-1) in 1763. The model made only a few turns before stopping. Watt realized that the small size caused an enormous loss of energy. Hence, he started to investigate the principles of the steam engine in his spare time, realizing in 1765 that the efficiency would improve significantly if the steam didn't condense inside of the cylinder but externally in a condenser. After additional researches on how to create a better seal inside of the cylinder, Watt developed an improved and more efficient engine in 1775 (Scherer 1965)[③]. The Watt engine (Figure 2-2) used only about one-quarter of the fuel of the Newcomen

---

① drain [dreɪn] *vt.* 使流出；排掉水

② ventilation [ˌventɪˈleɪʃ(ə)n] *n.* 通风设备；空气流通

③ Stating this in such few sentences makes it look obvious, but this does not give justice to Watt, who labored endlessly, calculating and researching uncountable details on the thermodynamics of steam and the principles of the steam engine, and trying out numerous technical approaches and solutions before achieving his breakthrough. Like most inventors, a great insight is usually preceded by a lot of hard work.

steam engine for the same work, making the former much more preferable over the latter①.

Figure 2-1 James Watt, by Henry Howard, ca. 1797

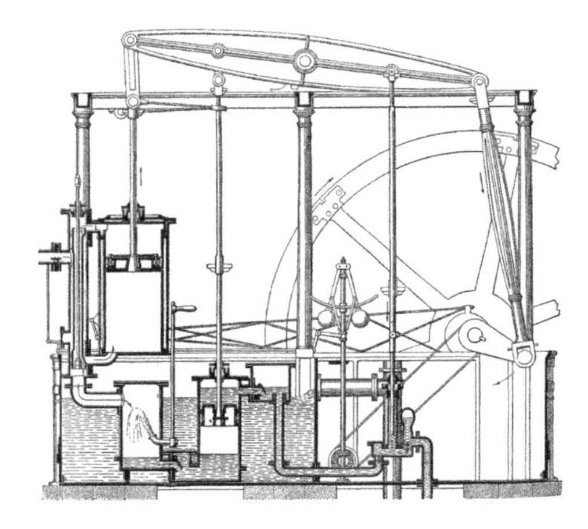

Figure 2-2 Steam Engine Designed by James Watt (Image from Meyers Konversations-Lexikon 1885-1890.)

While the major change during the industrial revolution was steam power, most mechanization in manufacturing was still powered by water. Steam power was expensive in installation, also required a constant supply of expensive fuel, needed highly trained experts to run the engine, and nevertheless frequently broke down, even if the risk of explosion was (mostly) under control. The advantage of the steam engine was that it could be installed almost anywhere, whereas waterpower required a constant supply of water. Up to 1780, almost all steam engines were used to pump water out of mines. Since the location of the mine was often not near a readily available water source, steam power was expensive but often the only viable option to pump water out of mines. Additionally, since steam engines needed lots of fuel to run, pretty much the only place to run them economically was next to a coal mine. Other locations incurred significant transport cost on coal, making the use of Newcomen steam engines too expensive compared to the use of horses or water. If the mine was near a suitable water source, however, waterpower was usually preferred, even with the risk of flooding the mine. Waterwheels were even installed inside of mines, for example, the Schwazer silver mine in Austria.

Yet, due to its expense, most manufacturing until 1870 was powered by

---

① There were also a large number of other tinkers working on steam engines such as Jonathan Hornblower, Francis Thompson, Richard Trevithick and many more. While they all contributed to the body of knowledge about steam engines, they were of minor commercial significance.

water, and waterwheels were commonly used until the mid-nineteenth century. In some cases, a steam engine was even installed with the sole purpose to pump water to drive a waterwheel, which provided much smoother motion than any contemporary steam engines. However, the Watt steam engine was initially able to convert only 5% of the heat in movement. Additionally, the waterwheel had an efficiency of less than 50%. Hence, the overall efficiency of the coupled system was probably no more than 2%.

For manufacturing purposes, steam engines were insignificant before 1780. Besides their higher fixed and variable costs and lower reliability, one additional limitation was that they provided only a back-and-forth[①] motion, whereas waterwheels provided a much more useful rotary motion. Only in 1779 did James Pickard modify a steam engine to provide a rotary motion, and from around 1785 onward most steam engines were rotary engines. While many of the rotary engines were also used in mining, from 1780 onward, more and more steam engines were used to drive manufacturing-related machinery. Steam engines were used in textile mills, to turn machinery, to pump air in a blast furnace, to grind corn and to roll metal. Probably one of the earliest large steam-powered factories was the Albion Flour Mills in London, planned to have 30 millstones and 3 steam engines. In comparison, the next largest mill in London had 4 millstones. However, local millwrights were strictly against the new competition, and the mill burned down in 1791 after only a few months of operation. Overall, about 450 steam engines or about 20% of all steam engines during the eighteenth century were used in manufacturing in Britain. Waterpower did generate roughly the same power as steam engines of about 15 to 30 horsepower[②] per installation. Some larger installations both in water and in steam exceeded 100 horsepower. Steam engines were mainly used in locations where waterpower was unavailable or unreliable, or by early adopters who were interested in new technology despite its risks.

As the technology improved, efficiency and power output went up, and the price went down. More and more steam engines were installed. By 1850, there were about 10 000 steam engines in Great Britain, with efficiency improving to 15% and an average power output of around 50 horsepower. The number of steam engines was still minor compared to downloaded by the up to 30 000 waterwheels for the same time and regions. The rest of Europe probably had another 10 000

① back-and-forth 反复，来回

② horsepower: a unit to measure the power of an engine.

steam engines installed. Waterpower also improved, and around 1860, large wheels provided up to 250 horsepower, but the largest engines quadrupled that with up to 1 000 horsepower. Around 1870, steam engines became more prominent than waterpower, and around 1900, the largest steam engine provided 5 000 horsepower. Finally, after the invention of the steam turbine by Charles Algernon Parsons in 1884, steam power was able to provide up to 130 000 horsepower. A single modern steam turbine can provide up to 2 000 000 horsepower, probably more than all medieval① waterwheels combined. In fact, most of our electricity is generated with the help of steam turbines, the great-great-grandchild of the Savery, Newcomen and Watt steam engines.

2.2 The History of the Design Process

During design activities, ideas are developed into hardware that is usable as a product. Whether this piece of hardware is a bookshelf or a space station, it is the result of a process that combines people and their knowledge, tools and skills to develop a new creation. This task requires their time and costs money, and if people are good at what they do and the environment they work in is well structured, they can do it efficiently. Further, if they are skilled, the final product will be well liked by those who use it and work with it—the customers will consider it as a quality product. The design process, then, is the organization and management of people and the information they develop in the evolution of a product.

In simpler times, one person could design and manufacture an entire product. Even for a large project such as the design of a ship or a bridge, one person might have sufficient knowledge of the physics, materials, and manufacturing processes to manage all aspects of the design and construction of the project.

By the middle of the twentieth century, products and manufacturing processes had become so complex that one person no longer had sufficient knowledge or time to focus on all the aspects of the evolving② project. Different groups of people became responsible for making, design, manufacturing and overall management. This evolution led to what is commonly known as the "over-the-wall" design process (Figure 2-3).

---

① medieval [ˌmedi'iːvl] *adj.* 中世纪的；原始的；仿中世纪的；老式的

② evolve [i'vɒlv] *vt.* & *vi.* 演变；进化

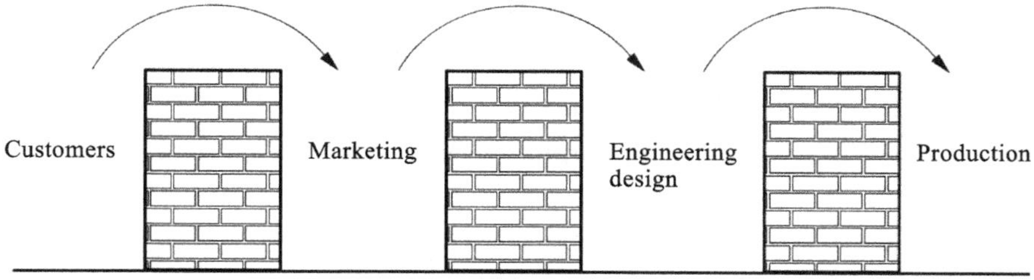

Figure 2-3 The Over-the-wall Design Process

In the structure shown in Figure 2-3, the engineering design process is walled off from other product development functions. Basically, people in marketing communicate a perceived① market need to engineers either as a simple, written request or, in many instances, orally②. This is effectively a one-way communication and is thus represented as information that is "thrown over the wall". Engineer interprets③ the request, develops concepts, and refines the best concept into manufacturing specifications④(i. e. drawings, bills of materials and assembly instructions). These manufacturing specifications are shown over the wall to be produced. Manufacturer then interprets the information passed to him and builds what he thinks the engineer wanted.

Unfortunately, often what is manufactured by a company using the over-the-wall process is not what the customer had in mind. This is because of the many weaknesses in this product development process. First, people in marketing may not be able to communicate to engineers a clear picture of what the customers want. Since the design engineers have no contact with the customers and limited communication with marketing people, there is much room for poor understanding of design problems. Second, design engineers do not know as much about the manufacturing process as manufacturing specialists, and therefore some parts may not be able to be manufactured as drawn or manufactured on existing equipment. Further, manufacturing experts may know less-expensive methods to produce the product. Thus, this single-direction over-the-wall approach is inefficient and costly and may result in poor-quality products. Although many companies still use this method, most are realizing its weaknesses, and are moving away from its use.

① perceive [pə'siːv] *vt.* 感觉，察觉，理解
② orally ['ɔːrəli] *adv.* 口头上地，口述地
③ interpret [ɪn'tɜːprɪt] *vt.* 解释；说明
④ manufacturing specification 制造说明书；制造技术条件

In the late 1970s and early 1980s, the concept of simultaneous engineering[①] began to break down the walls. This philosophy emphasized the simultaneous development of the manufacturing process with the evolution of the product. Simultaneous engineering was accomplished by assigning manufacturing representatives to be members of design teams so that they could interact with the design engineers throughout the design process. The goal was the simultaneous development of the product and the manufacturing process.

In the 1980s, the simultaneous design philosophy was broadened and called concurrent engineering[②], which, in the 1990s, became Integrated Product and Process Design (IPPD). Although the terms simultaneous, concurrent and integrated are basically synonymous[③], the change in terms implies a greater refinement in thought about what it takes to efficiently develop a product. Throughout the rest of this text, the term concurrent engineering will be used to express this refinement.

In the 1990s, the concepts of Lean and Six Sigma became popular in manufacturing and began to have an influence on design. Lean manufacturing[④] concepts were based on studies of the Toyota manufacturing system and introduced in the United States in the early 1990s. Lean manufacturing seeks to eliminate waste in all parts of the system, principally through teamwork. This means eliminating products nobody wants, unneeded steps, many different materials, and people waiting downstream because upstream activities haven't been delivered on time. In design and manufacturing, the term "lean" has become synonymous with minimizing the time to do a task and the material to make a product.

Lean focuses on time, while Six Sigma focuses on quality. Six Sigma, sometimes written as "6σ", was developed at Motorola in the 1980s and popularized in the 1990s as a way to help ensure that products were manufactured to the highest standards of quality. Six Sigma uses statistical methods to account for and manage product manufacturing uncertainty and variation. Key to Six Sigma methodology is the five-step DMAIC process (define, measure, analyze, improve, and control). Six Sigma brought improved quality to manufactured products. However, quality begins in the design of products, and processes, not in their manufacturing.

---

① simultaneous engineering 同步工程，同时工程

② concurrent engineering 并行工程

③ synonymous [sɪ'nɒnɪməs] *adj.* 同义词的，同义的，意思相同的

④ lean manufacturing 精益制造

Recognizing this, the Six Sigma community began to emphasize quality earlier in the product development cycle, evolving DFSS (Design for Six Sigma) in the late 1990s.

Essentially DFSS is a collection of design best practices similar to others. DFSS is still an emerging discipline.

Beyond these formal methodologies, during the 1980s and 1990s many design process techniques were introduced and became popular. They are essential building blocks of the design philosophy.

All of these methodologies and best practices are built around a concern for the ten key features listed in Table 2-1. The primary focus is on the integration of teams of people, design tools and techniques, and information about the product and the processes used to develop and manufacture it.

Table 2-1 The Ten Key Features of Design Best Practice

1. Focus on the entire product life
2. Use and support of design teams
3. Realization that the processes are as important as the product
4. Attention to planning for information-centered tasks
5. Careful product requirements development
6. Encouragement of multiple concept generation and evaluation
7. Awareness of the decision-making process
8. Attention to designing in quality during every phase of the design process
9. Concurrent development of product and manufacturing process
10. Emphasis on communication of the right information to the right people at the right time

The use of team, including all the "stakeholders" (people who have a concern for the product), eliminates many of the problems with the over-the-wall method. During each phase in the development of a product, different people will be important and will be included in the product development team. This mix of people with different views will also help the team address the entire life cycle of the product.

Tools and techniques connect the teams with the information. Although many of the tools are computer-based, much design work is still done with pencils and paper.

Chapter 3 The Way to Read and Understand Professional Literature

Reading is not only concerned about how many words and sentences, but more importantly, what knowledge has been learned. First read, and then understand, and then use.

In this chapter, different areas of mechanical design and manufacturing will be used as text examples. By exercises after texts, it will help you to form your method to understand how to read and understand professional literature in some extent.

3.1 Reading Experience Ⅰ: Developing Vocabulary Exercises

3.1.1 Picking out the Professional Words

Each occupational area has its own language. Before you can comprehend the "plain sense" of the instructional materials you are to read, you have to understand the language used. A precise, working understanding of hundreds of difficult terms is often crucial.

For example, not knowing the difference between the terms *boring* and *counterboring* can ruin a project in a woodworking shop. Since knowledge of technical terminology is so important in all vocational-technical programs, vocabulary study should be an integral part of reading assignments. Significant terms are usually associated with the key concepts to be learned, the important processes to be performed, and the tools or instruments to be used. For this reason, the identification of major concepts/processes/tools and key vocabulary words should be an important part of the planning you do to learn reading.

You will probably have little trouble in picking out the professional words to be learned. You need to look for the following.

- New words that have not appeared in your reading previously
- New technical terms

- Complex or compound terms
- Everyday words that have a special meaning in your occupational area
- Key words that must be known in order to understand the particular topic

TOPIC 1: Engineering Materials

Selection of Materials

An increasingly wide variety of materials are now available and each type has its own materials and properties and manufacturing characteristics, advantages and limitations, material and production costs, and consumers and industrial applications. The selection of materials for products and their components is typically made in consultation with material engineers, although design engineers may also be sufficiently experienced and qualified to do so. At the forefront of new material usage are industries such as the aerospace and aircraft, automotive, military equipment and sporting goods industries.

The general types of materials used, either individually or in combination with other materials, are the following.

(1) Ferrous① metals: carbon, alloy, stainless, and tool die steels.

(2) Nonferrous metals: aluminum, magnesium, copper, nickel, titanium, superalloys, refractory metals, beryllium, zirconium, low-melting-point alloys and precious metals②.

(3) Plastics (polymers): thermoplastics, thermosets and elastomers③.

(4) Ceramics, glasses, glass ceramics, graphite, diamond and diamond-like materials.

(5) Composite materials: reinforced plastics and metal-matrix and ceramic-matrix composites.

(6) Nanomaterials

(7) Shape-memory alloys (also called smart materials), amorphous alloys, semiconductors and superconductors④.

As new developments continue, the selection of an appropriate material for a particular application becomes even more challenging. Also, there are continuously

---

① ferrous ['ferəs] *adj.* 铁的，含铁的

② aluminum, magnesium, copper, nickel, titanium, superalloy, refractory metal, beryllium, zirconium, low-melting-point alloy and precious metal 铝、镁、铜、镍、钛、高温合金/高耐热合金、难熔金属、铍、锆、低熔点合金、贵金属

③ thermoplastic, thermoset, and elastomer 热塑性塑料、热固性塑料、人造橡胶

④ amorphous alloy, semiconductor and superconductor 非晶合金、半导体、超导体

shifting trends in the substitution of materials, driven not only by technological considerations, but also economics.

Properties of materials. *Mechanical properties* of interest in manufacturing generally include strength, ductility, hardness, toughness, elasticity, fatigue and creep resistance. *Physical properties* are density, specific heat, thermal expansion and conductivity, melting point, and electrical and magnetic properties. Optimum designs often require a consideration of a combination of mechanical and physical properties. A typical example is the strength-to-weight and stiffness-to-weight ratio of materials for minimizing the weight of structural members. Weight minimization is particularly important for aerospace and automotive applications, in order to improve performance and fuel economy.

Chemical properties include oxidation, corrosion, degradation, toxicity① and flammability②. These properties play a significant role under both hostile (such as corrosive) and normal environments. *Manufacturing properties* indicate whether a particular material can be cast, formed, machined, joined, and heat treated with relative ease. As Table 3-1 illustrates, no one material has the same manufacturing characteristics. Another consideration is *appearance*, which includes such characteristics as color, surface texture and feel, all of which can play a significant role in a product's acceptance by the public.

Availability. The economic aspect of material selection is as important as technological considerations. Thus, the availability of materials is a major concern in manufacturing. Furthermore, if materials are not available in the desired quantities, shapes, dimensions and surface texture, substitute materials or additional processing of a particular material may well be required, all of which can contribute significantly to product cost.

Reliability of supply is important in order to meet production schedules. In automotive industries, for example, materials must arrive at a plant at appropriate time intervals. Reliability of supply is also important, considering the fact that most countries import numerous raw materials. The United States, for example, imports most of the cobalt, titanium, chromium③, aluminum, nickel, natural rubber, and diamond that it needs. Consequently, a country's *self-reliance* on resources, especially energy, is an often-expressed political goal, but is challenging

---

① toxicity [tɒk'sɪsəti] *n.* 毒性

② flammability [ˌflæmə'bɪlətɪ] *n.* 易燃性

③ cobalt, titanium, chromium 钴、钛、铬

to achieve. Geopolitics① (defined briefly as the study of the influence of a nation's physical geography on its foreign policy) must thus be a consideration, particularly during periods of global hostility.

Table 3-1 General Manufacturing Characteristics of Various Materials

Alloy	Castability	Weldability	Machinability
Aluminum	E	F	E-G
Copper	G-F	F	G-F
Cray cast iron	E	D	G
White cast iron	G	VP	VP
Nickel	F	F	F
Steels	F	E	F
Zinc	E	D	E

Note: E, excellent; G, good; F, fair; D, difficult; VP, very poor. The ratings shown depend greatly on the particular materials, its alloys, and its processing history.

Service life. We all have had the experience of a shortened service life of a product, which often can be traced to improper selection of materials, improper selection of production methods, insufficient control of processing variables, defective parts or manufacturing-induced defects, poor maintenance, and improper use of the product. Generally, a product is considered to have failed when the following situations occur to it.

(1) Stops functioning, due to the failure of one or more of its components, such as a broken shaft, gear, bolt, or turbine blade② or a burned-out electric motor.

(2) Does not function properly or perform within required specifications, due, for example, to worn gears or bearings.

(3) Becomes unreliable or unsafe for further use, as in the erratic③ behavior of a switch, poor connections in a printed-circuit board, or delamination④ of a composite material.

This text describes the types of failure of a component or a product resulting, for example, ① design deficiencies, ② improper material selection, ③ incompatibility of

① geopolitics [ˌdʒiːəʊˈpɒlətɪks] *n.* 地理政治论，地缘政治学
② shaft, gear, bolt, or turbine blade 轴，齿轮，螺栓或涡轮叶片
③ erratic [ɪˈrætɪk] *adj.* 不稳定的，无规律的
④ delamination [diːˌlæmɪˈneɪʃən] *n.* 分层，剥离

materials in contact, which produces friction, wear, and galvanic[①] corrosion, ④ defects in raw materials, ⑤ defects induced during manufacturing, ⑥ improper component assembly, and ⑦ improper product use.

Material substitution in products. For a variety of reasons, numerous substitutions are often made in materials, as evidenced by a simple inspection and comparison of common products such as home appliances, sports equipment, or automobiles. As a measure of the challenges faced in material substitution, consider the following examples: ① metal vs. wooden handle for a hammer, ② aluminum vs. cast-iron lawn chair, ③ aluminum vs. copper wire, ④ plastic vs. steel car bumper, ⑤ plastic vs. metal toy, and ⑥ alloy steel vs. titanium submarine hull.

The following two examples in material substitution give typical details of the major factors involved in common products.

Example 1

Baseball Bats

Baseball bats for the major and minor leagues are generally made of wood from the northern white ash tree, a wood that has high dimensional stability, a high elastic modulus and strength-to-weight ratio, and high shock resistance. Wooden bats can, however, break during their use and may cause serious injury. The wooden bats are made on semiautomatic lathes and then subjected to finishing operations for appearance and labeling. The straight uniform grain required for such bats has become increasingly difficult to find, particularly when the best wood comes from ash trees that are at least 45 years old.

For the amateur market and for high school and college players, aluminum bats have been made since the 1970s as a cost-saving alternative to wood. The bats are made by various metalworking operations. Although, at first, their performance was not as good as that of wooden bats, the technology has advanced to a great extent. Metal bats are now made mostly from high-strength aluminum tubing, as well as other metal alloys. The bats are designed to have the same center of percussion (known as the sweet spot, as in tennis racquets) as wooden bats, and are usually filled with polyurethane or cork for improved sound damping and for controlling the balance of the bat.

Metal bats possess such desirable performance characteristics as lower weight than wooden bats, optimum weight distribution along the bat's length, and superior impact dynamics. Also, as documented by scientific studies, there is a

① galvanic [gælˈvænɪk] *adj.* 以化学方法产生电流的

general consensus that metal bats outperform wooden bats.

Development in bat materials now includes composite materials consisting of high-strength graphite and glass fiber embedded in and epoxy resin matrix. The inner woven sleeve is made of Kevlar fibers (an aramid), which add strength to the bat and dampen its vibrations. These bats perform and sound much like wooden bats.

Source: Mizuno Sport, Inc.

Example 2

U. S. Pennies

Billions of pennies are produced and put into circulation each year by the U. S. Mint. The materials used have undergone significant changes throughout history, largely because of periodic material shortages and the resulting fluctuating cost of appropriate raw materials. The following shows the chronological development of material substitutions in pennies.

1793-1837	100% cooper
1837-1857	95% copper, 5% tin and zinc
1857-1863	88% copper, 12% nickel
1864-1962	95% copper, 5% tin and zinc
1943 (World war Ⅱ)	Steel, plated with zinc
1962-1982	95% copper, 5% zinc
1982-present	97.5% zinc, plated with zinc

EXERCISE 1: Picking out the Professional Words

Please list the technical terminology in WORK SHEET after reading TOPIC 1.

WORK SHEET

	Selection of Materials
1. New words that have not appeared in your reading previously	1. 2. 3. 4. 5.
2. New technical terms	1. 2. 3. 4. 5.

Continued WORK SHEET

	Selection of Materials
3. Complex or compound terms	1. 2. 3. 4. 5.
4. Everyday words that have a special meaning in your occupational area	1. 2. 3. 4. 5.
5. Key words that must be known in order to understand the particular topic	1. 2. 3. 4. 5.

3.1.2 Types of Context Clues

Most reading specialists agree that the single most important vocabulary skill is the use of context clues—the ability to search out a word's meaning from clues given by other words that surround it. Little children acquire vocabulary in this way—by hearing words used over and over again in similar situations or contexts. Adults continue to learn in this fashion.

For this reason, it may be difficult to give a precise meaning for a word we know. When asked to define a particular word, we may respond, "I can't give you a definition, but I can use it in a sentence." The way in which the word is used in the sentence gives some ideas of its meaning—again, through its place in the sentence and the meaning of words preceding and following it.

To use context clues effectively, readers must first recognize the fact that the context can provide clues to the meaning of an unknown word. They must also keep in mind that the context may reveal the meaning of a word only partially. In order to be helpful for the clues, they should be near the word, preferably within the same sentence or paragraph.

Types of Context Clues

Context clues can be grouped in six categories: outright definitions, examples, modifiers, restatement, inference, and inference through established connections. It is not necessarily important that you memorize the label for each of these categories. However, it is important that you know how to recognize each type of clues and understand how each reveals the meaning of a word.

Outright definition. This is the easiest context clue to use, since the purpose of the sentence is to give a direct definition. The usual pattern is as follows: the unknown word + a form of the verb *to be* + a definition. For example:

Cerebral hemorrhage is another name for stroke.

In this sentence, the technical term *cerebral hemorrhage* is linked by the word *is*, which is like an equal sign to the word *stroke*. Rewritten, it could read as follows:

Cerebral hemorrhage = Stroke

You need to recognize the words that are often used as equal signs: *means*, *can be defined as*, *called*, *termed*, and so forth.

Example. Examples of commonly known things, with which the reader is likely to be familiar, are also frequently used as context clues. They are given to help the reader understand a more general term with which he/she may not be familiar. Signal words (*e. g. like*, *such as*, *for example*) are often used with examples. For instance:

Many *legume* vegetables, such as navy beans, soybeans, peas, and lentils, can be dried and stored for long periods.

This sentence tells the reader that navy beans, peas, and so on are legume vegetables, helping the reader to understand the possibly unfamiliar term *legume*.

Modifier. Modifiers may be phrases, clauses, or single words, often in the form of predicate adjectives. They are intended to give a more precise meaning of the word they modify. For example:

To cut curves in thin wood, one should use a thin-bladed, fine-toothed *scroll saw*.

In this sentence, the phrase *thin-bladed*, *fine-toothed* modifies, or gives a more precise description of the term *scroll saw*.

Restatement. A restatement is announced by signal words such as *that is to say*, *that is*, *in other words*, *what this means*, *or to put it another way*. A restatement may also be announced using the word *or*, followed by a synonym. Sometimes dashes or parentheses indicate a restatement. For example:

Agglutinins are chemicals that *agglutinate* cells—that is, make them stick together in clumps.

In this example, the restatement tells us that the word *agglutinate* means to make something stick together.

Inference. Inference is the process of gathering details and "reading between the lines" in order to perceive relationships that have not been explicitly stated. In other words, no signal words are present to connect the term with an explanation of its meaning. However, by using *reason*, *logic*, and *speculation* (in short, inference) you can deduce such an explanation. For example:

The welding operation should be shielded so that no one in the vicinity may be in a position to look directly at the arc or have it shine in his/her eyes. If someone should accidentally become severely *flashed*, special treatment should be given at once by a physician.

In this case, the meaning of the word *flashed* can be inferred from two clues. One: The first sentence shows that it has something to do with looking directly at the extremely bright light of the welding arc. Two: The fact that a person looking at an arc may need treatment by a physician is an obvious clue that being flashed is hazardous and can cause severe eye damage.

Inference through established connections. This context clue depends on relationships established by sentence construction—repetition of key words or the use of connecting words that indicate comparison or contrast. For example:

What lay people call "strokes" or "apoplexy," physicians call "cerebrovascular accidents."

Teak wood has many of the same uses as black walnut but is harder to work, lighter in color, and close-grained and oily rather than open-grained.

In the first example, the common terms and the technical terms are linked by key words: "What lay people call... physicians call..." In the second example, connecting words indicate how teak contrasts with black walnut: "... *but* is harder... lighter... close-grained and oily *rather than* open-grained."

The procedure of recognizing contexts clues is as follows.

(1) Identify any signal words in order to determine what kind of information the contests providing.

(2) Point out the clue section of the sentence.

(3) Analyze the clue section to identify possible meanings of the unknown word.

(4) List some possible meanings from which the students can select the best or correct meaning.

TOPIC 2: Mechanisms

Clutch① Mechanisms

A *clutch* is defined as a coupling that connects and disconnects the driving and driven parts of a machine; an example is an engine and a transmission. A clutch typically contains a driving shaft and a driven shaft, and they are classed as either externally or internally controlled. Externally controlled clutches can be controlled either by friction surfaces or components that engage or mesh positively. Internally controlled clutches are controlled by internal mechanisms or devices; they are further classified as *overload*, *overriding*, and *centrifugal*. There are many different schemes for a driving shaft to engage a driven shaft.

Externally Controlled Friction Clutches

Friction-Plate Clutch②. This clutch has a control arm, which when actuated, advances a sliding plate on the driving shaft to engage a mating rotating friction plate on the same shaft; this motion engages associated gearing that drives the driven shaft. When reversed, the control arm disengages the sliding plate. The friction surface can be on either plate, but is typically only on one. As shown in Figure 3-1, When the left sliding plate on the driving shaft is clamped by the control arm against the right friction plate idling on the driving shaft, friction transfers the power of the driving shaft to the friction plate. Gear teeth on the friction plate mesh with a gear mounted on the driven shaft to complete the transfer of power to the driven mechanism. Clutch torque depends on the axial force exerted by the control arm.

Cone Clutch③. This clutch operates on the same principle as the friction-plate clutch except that the control arm advances a cone on the driving shaft to engage a mating rotating friction cone on the same shaft; this motion also engages any associated gearing that drives the driven shaft. The friction surface can be on either cone but is typically only on the sliding cone.

Expanding Shoe Clutch④. This clutch is similar to the friction-plate clutch except that the control arm engages linkage that forces several friction shoes radially outward so they engage the inner surface of a drum on or geared to the

---

① clutch [klʌtʃ] *n.* 离合器

② friction-plate clutch 摩擦式离合器，摩擦板离合器

③ cone clutch 锥形离合器

④ expanding shoe clutch 胀瓦式离合器

driven shaft.

Externally Controlled Positive Clutches

Jaw Clutch①. This clutch is similar to the friction-plate clutch except that the control arm advances a sliding jaw on the driving shaft to make positive engagement with a mating jaw on the driving shaft.

Other examples of externally controlled positive clutches are the *planetary transmission clutch* consisting essentially of a sun gear keyed to a driving shaft, two planet gears, and an outer driven ring gear; the *pawl and ratchet clutch* consists essentially of a pawl-controlled driving ratchet keyed to a driven gear.

Internally Controlled Clutches

Internally controlled clutches can be controlled by spring, torque, or centrifugal force. The *spring and ball radial-detent clutch*, for example, disengages when torque becomes excessive, allowing the driving gear to continue rotating while the driving shaft stops rotating. The *wrapped-spring clutch* consists of two separate rotating hubs joined by a coil spring. When driven in the right direction, the spring tightens around the hubs increasing the friction grip. However, if driven in the opposite direction, the spring relaxes, allowing the clutch to slip.

The *expanding-shoe centrifugal clutch* is similar to the externally controlled *expanding shoe clutch* except that the friction shoes are pulled in by springs until the driving shaft attains a pre-set speed. At that speed centrifugal force drives the shoes radially outward so that they contact the drum. As the driving shaft rotates faster, pressure between the shoes and drum increases, thus increasing clutch torque.

The *overrunning or overriding clutch*, as shown in Figure 3-2, is a specialized form of a cam mechanism, also called a *cam and roller clutch*. The inner driving cam A has wedge-shaped notches② on its outer rim that hold rollers between the outer surface of driving cam A and the inner cylindrical surface of outer driven ring B. When driving cam A is turning clockwise, frictional forces wedge the rollers tightly into the notches to lock outer driven ring B in position so it also turns in a clockwise direction. However, if driven ring B is reversed or runs faster clockwise than driving cam A (when it is either moving or immobile), the rollers are set free, the clutch will slip and no torque is transmitted. Some versions of this clutch include springs between the cam faces and the rollers to ensure faster clutching

---

① jaw clutch 爪式离合器；颚夹离合器；牙嵌离合器

② notch [nɒtʃ] *n.* 缺口，槽口

action if driven ring B attempts to drive driving cam A by overcoming residual friction. A version of this clutch is the basic free-wheel mechanism that drives the rear axle of a bicycle.

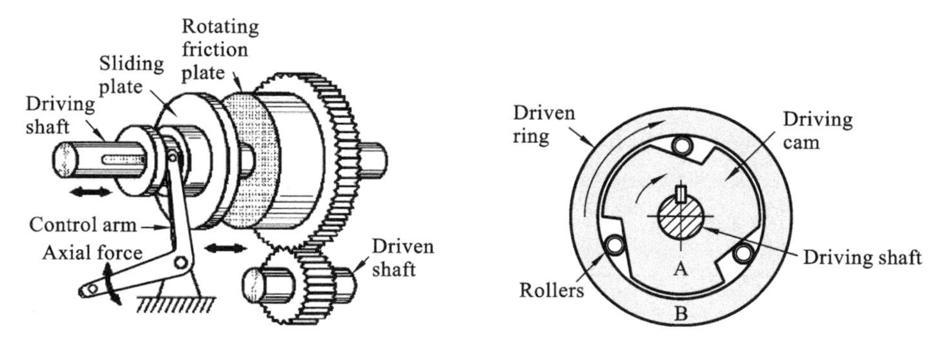

Figure 3-1 Friction-Plate Clutch

Figure 3-2 Overrunning Clutch

Some low-cost, light-duty overrunning clutches for one-direction-only torque transmission intersperse cardioid[①]-shaped pellets called sprags with cylindrical rollers. This design permits cylindrical internal drivers to replace cammed drivers. The sprags bind in the concentric space between the inner driver and the outer driven ring if the ring attempts to drive the driver. The torque rating of the clutch depends on the number of sprags installed. For acceptable performance a minimum of three sprags, equally spaced around the circumference of the races, is usually necessary.

EXERCISE 2: Types of Context Clues

Directions: By reading materials in TOPIC 2, it helps to know how to search out the meanings of new word from clues given in the sentence. Read the following sentence, and then answer the questions. Circle the best responses to items (1)-(4). Write out your response to item (5).

A clutch is defined as a coupling that connects and disconnects the driving and driven parts of a machine; an example is an engine and a transmission.

(1) What word are we trying to define?

a. coupling b. clutch c. driving and driven parts

(2) What words in the sentence signal you that some information will follow to tell you about this signal?

a. example b. engine c. transmission

① cardioid [ˈkɑːdɪɔɪd] *n.* 心脏形曲线

(3) What kind of information do you expect to follow this signal?

a. opposites　　b. samples　　c. restatement

(4) According to the sentence, what does an engine and a transmission have in common?

a. They are mechanical parts.

b. They are connected with each other.

c. They consist of driving and driven parts.

(5) If you work as a mechanical design engineer, what does this sentence tell you about selecting clutch?

3.1.3 Context Clues in Specialized Vocabulary

Context clues are often essential in finding the meanings of specialized vocabulary words. Specialized words are those words that have common general meanings, but have different and highly specific meanings when used in a given subject area.

Take the word *radical* as an example. To a nursing instructor, radical surgery means the most extensive surgery available for the condition. To a mathematics teacher, radical pertains to a root. To a social studies teacher, it means favoring drastic political or social ideas.

There are many such specialized words in vocational education. In auto mechanics, the following are examples of everyday words with specialized meanings: bleeding, bounce, shimmy, race, and tramp. In machine shop, such ordinary words as backlash, female, journal, and worm take on special meanings.

TOPIC 3: Mechanical Design

Use Three-Dimensional Solid Model Layouts to Find the Best Arrangement of Part and Assemblies

Engineers and designers use graphical representations for two important purposes: ① as an aid to thinking, ② for communicating design ideas.

There are two traditional types of graphical representation: orthographic[①]

---

① orthographic [ˌɔːθəˈgræfɪk] *adj.* 正交的

projection① and perspective drawing②. Three-dimensional (3D) solid models have become a third type of graphical representation, and have largely replaced the other two for designing mechanical assemblies, both as a thinking and a communication tool, as shown in Figure 3-3, a 3D fully-dimensioned model captures more information than the other two.

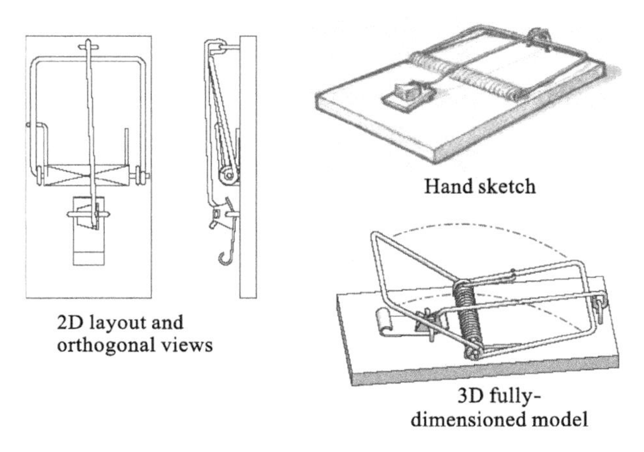

Figure 3-3 2D Layout, 3D Sketch, and 3D Model

Let's be truthful. 3D solid modeling is how mechanical design is done. Not every step of the way, and not by everybody, but it is the predominant③ tool for defining geometry during mechanical design. The advantages are overwhelming: exceptional visualization, continuity of effort, and simple, unambiguous④ communication.

So why has 3D modeling been embraced by the mechanical design community? For the same reasons that graphical representations have always been useful, but even more so. Everyone's visualization skills and memory, no matter how good, have a limit. Graphical representations supplement your short term memory and allow you to build upon what you already have. 3D solid models do this dramatically better, and you cannot afford to forgo⑤ this advantage.

To design a mechanical assembly, make a 3D layout, start with simple representations of the parts, assemble them together, and check for fit and function. If you haven't learned 3D modeling, work with someone who has. This

---

① projection [prə'dʒekʃən] *n.* 投影
② perspective drawing 透视图
③ predominant [prɪ'dɒmɪnəns] *adj.* 主要的；突出的；最显著的
④ unambiguous [ˌʌnæm'bɪgjuəs] *adj.* 不含糊的；清楚的；明确的
⑤ forgo [fɔː'gəʊ] *v.* 没有……也行；放弃

layout is an essential representation of a mechanical assembly during design. Why?

Scale. It is the rare designer who can grasp, without the aid of scale representation, all of the constraints①, limitations, and possibilities of a design concept. Often, a promising idea simply does not work because there is no room for all the parts and functions. Furthermore, few design projects start with completely clean sheets of paper. More often, we must design around—and fit into—what already exists. Most of us need help, and nowadays, 3D solid models provide it.

But there are caveats② and exceptions!

One obvious exception is simple parts. I will freehand draw and dimension a part if it has only a few features and it does not go into an assembly. (If it is part of an assembly, I probably need a 3D model anyway to check for fit and function.) Another is for parts with predominantly two-dimensional characteristics such as extrusion profiles, stampings③, or die④ cuts.

An accepted caveat is not to use 3D modeling for ideation⑤, brainstorming, and the like. Some do, some do not; this will depend upon your and your organization's preference. This text recommends using 3D modeling "to find the best arrangement of parts and assemblies." How initial concepts and multiple alternatives are identified, documented, and presented is something different.

Making 3D CAD models takes time, and is likely to be slower than your brain or a brainstorming group during the ideation stage of a project. So indeed, you may want to create hand sketches, both orthographic and perspective, especially for initial concepts, as shown in Figure 3-4. Sketches are valuable when generating multiple concepts. 3D CAD models are often unnecessary at the early ideation stage of a project.

There is an exception to this exception. You may have to use 3D layouts as part of the concept selection process. How can you decide among different designs without understanding their geometric constraints? I have more than once selected the "best" design among different options, only to find intractable⑥ space or size difficulties with a 3D layout, so this can be a vital part of the screening process.

You may prefer to sketch at least a little before starting a 3D model. A sketch

---

① constraint [kən'streɪnt] *n.* 约束

② caveat ['keiviæt] *n.* 警告；告诫

③ stamping[stæmpɪŋ] *n.* 冲压件

④ die 模具，冲模

⑤ ideation [ˌaɪdi'eɪʃən] *n.* 构思能力，思维能力；构思过程

⑥ intractable [ɪn'træktəbl] *adj.* 难控制的，难压制的；难处理的

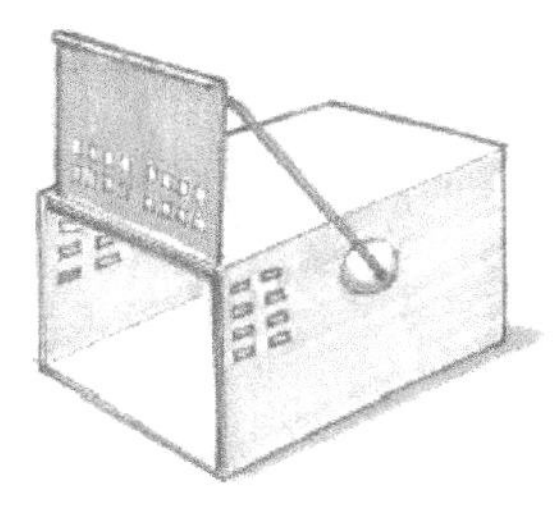
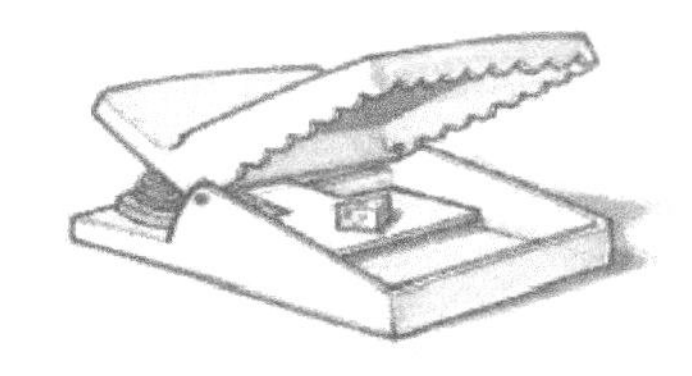
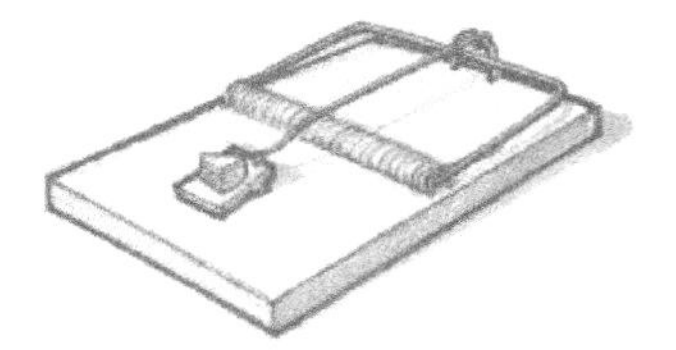

Figure 3-4 Sketches

can suggest a starting configuration, one that will prevent wasting 3D modeling time. But most designers of mechanical assemblies jump to 3D solid models quickly, sometimes right from their mind's eyes. Investing too much time in a hand sketch wastes time. You will need a 3D model before long anyway, so once you are committed to a direction, invest in the model instead.

But you must have discipline when creating 3D models. Do not create any more detail than is necessary at each stage—this wastes more time than you can afford. Functional prototypes do not need the same level of detail as a final molded part. Test it first, add detail later.

Finally, never let a design's history decide its future. Having invested a small career in a 3D model is not a sufficient reason to use it. "It is too much work to redo" will not justify a substandard design. If there is trouble, fix it. If you have to scrap it, scrap it. Chalk it up to experience, and don't make the same mistake next time.

EXERCISE 3: Everyday Words with Specialized Professional Meanings

The following worksheet shows some words that take on special meanings when used in mechanical design. Please find more words in TOPIC 3 and other materials related to the area of mechanical design.

WORK SHEET

	General meaning	Mechanical meaning
layout		
orthogonal view		

Continued WORK SHEET

	General meaning	Mechanical meaning
Scale		
constraint		
die		
stamp/stamping		
prototype		

3.1.4 Using the Word Structural Analysis to Understand Professional Words

Another important way to unlock the approximate meanings of words is by structural analysis, or studying the word parts. Some words can be divided into parts called roots, prefixes, and suffixes.

A **root** is the base or fundamental part of a word (e.g. *port*, which means "to carry"). A root can often stand alone as a word. Some words, by the way, are composed of two roots.

A **prefix** is a letter or sequence of letters that comes before a root. Its function is to change or modify the meaning of the root. As in the word *transport*, *trans* is a Latin prefix meaning "across". Thus, *transport* means to carry across. *Ex* is a Latin prefix meaning "out of". Thus, *export* means to carry out of.

A **suffix** comes after the root. A suffix may entirely change the meaning of a word, but its usual function is to change the part of speech of the base word. For example:

trans + port +*ed* = past tense of the verb *transport*

trans + port +*ation* = noun; the act of transporting

trans + port +*able* = adjective; capable of being transported

trans + port +*er* = noun; one who transports

Prefixes and suffixes are not usually derived from English words. Rather, they derive from other languages, such as Latin or Greek. A good dictionary is a great help in analyzing words, because it gives the meanings of prefixes and suffixes as well as whole words.

A knowledge of prefixes and suffixes is extremely helpful in figuring out the meanings of new words. Knowledge of such prefixes as *auto-*, *bio-*, *hydro-*, *lith-*, *photo-*, *pneumo-*, and others can reveal the meanings of hundreds of words, including the technical terms used in various occupational programs. Understanding such common suffixes as *-ability*, *-meter*, *-ology*, and *-chrome* can unlock the meanings of hundreds more.

Table 3-2 shows how the process of word analysis can be used to derive the meanings of words drawn from various technical fields. As can be seen in this table, there are limitations to the use of structural analysis. When a word is broken into several parts, and each part is defined with its Greek or Latin meaning, the result will be a literal translation. This may be quite different from the actual meaning of the word. For example, *photography* = photo + graph = light writing. While this analysis does hint at the process of photography. It is only a hint.

Table 3-2 Word Analysis

Analysis			Meaning
Tachometer	*Tacho*	=Speed	Device to measure speed of engine
	Meter	=Measuring instrument	
Chronoscope	*Chrono*	=Time	Device to measure very short lengths of time
	Scope	=Viewing instrument	
Vitrification	*Vitri*	=Glass	Conversion into glass
	Fication	=Made into	
Chromatic	*Chrom*	=Color	Pertaining to color
	Atic	=Pertaining to	

Continued Table 3-2

Analysis			Meaning
Hydrofoil	*Hydro*	=Water	Boat with winglike part that lifts it in the water
	Foil	=Wing shapes	
Isometric	*Iso*	=Equal	Perspective drawing with all dimensions to scale
	Metric	=Measure	

Such an oblique hint might be confusing if you were trying to figure out the meaning of a totally unknown word. In cases like that, you may first need to know the definition of the new word. Then, you could use your knowledge of word parts to help you grasp the meaning of the word, and as a device for remembering the meaning of the word. In that way, technical terms should begin to make a lot more sense to you.

Another problem with structural analysis is that some prefixes have two or more often unrelated meanings. For example, *in-* means "not" in such words as *indirect* and *ineffective*. It means "into" in such words as *incision* and *insight*.

A third problem with structural analysis is that the initial letters in some words look like prefixes while they aren't, as in the words *equipage* and *equine*. *Equi* is not a prefix meaning "equal".

Despite these limitations, structural analysis can be a useful method for figuring out word meanings. This skill, like that of using context clues, involves a problem-solving approach to word study.

TOPIC 4: Mechanical Equipments and Tools

Computer Numerical Control Machine Tools—Machining Centers, Milling Machines

Over the years, the influence of numerical controllers on the construction of machine tools has led in some cases to completely new machines and mechanical automation equipment. Today, computer numerical control (CNC) machines are the basic cornerstones of modern manufacturing equipment.

The various types of machines will be discussed below in descending order of their importance on the market. This is based on sales figures from the German Machine Tool Builders' Association.

Machining centers are machine tools that arose only as a result of the development of numerical control (NC) systems. They were developed from machine tools with rotating drives, that is, drilling machines, milling machines, or boring mills[①]. The goal was to be able to perform as wide a range of machining operations as possible automatically in a single setup. The definition therefore is as follows.

A machining center is a machine tool that is numerically controlled in at least three axes and is equipped with an automatic tool-changing system and a tool magazine.

If the machine is also capable of machining with a rotating workpiece, then it also can be called a turning center or turning/milling center.

There are many possible variants for machining centers. First of all, one can make a distinction between horizontal and vertical machines based on the position of the main spindle. Vertical machines are preferred for the machining of flat sheets or very long workpieces, whereas horizontal machines are used more for machining of box-shaped workpieces.

A distinction is made based on the number of feed axes.

(1) Three-axis machines: three linear axes, it is the most basic configuration for a machine with a rotating tool.

(2) Four-axis machines: three linear axes and one rotary axis; the purpose of the rotary axis is to allow machining on all sides. For horizontal machines, this is in the form of a rotary table, and for vertical machines, it is in the form of a reversible clamping device[②] for the machining of cylinder barrel surfaces (Figure 3-5 (b)) or for the machining of small workpieces on three sides (Figure 3-5 (a)).

(3) Five-axis machines: three linear axes and two rotary axes; this allows the tool to be moved in any desired directions relative to the workpiece. This means that milling is possible in any plane regardless of its orientation in space, and holes can be drilled in any diagonal orientation. The two rotary axes can be assigned as desired to the workpiece holder and the tool spindle. As a result, a large number of different machine configurations are possible.

1. Three-Axis Machines

Three-axis machines with three linear axes are the most basic configuration for machining centers. The implemented constructions as a vertical machine are shown in Figure 3-6.

2. Four-Axis Machining Centers

These generally consist of three linear CNC axes and a rotary table, making it

---

① boring mill 镗床

② clamping device 夹紧机构，夹紧装置

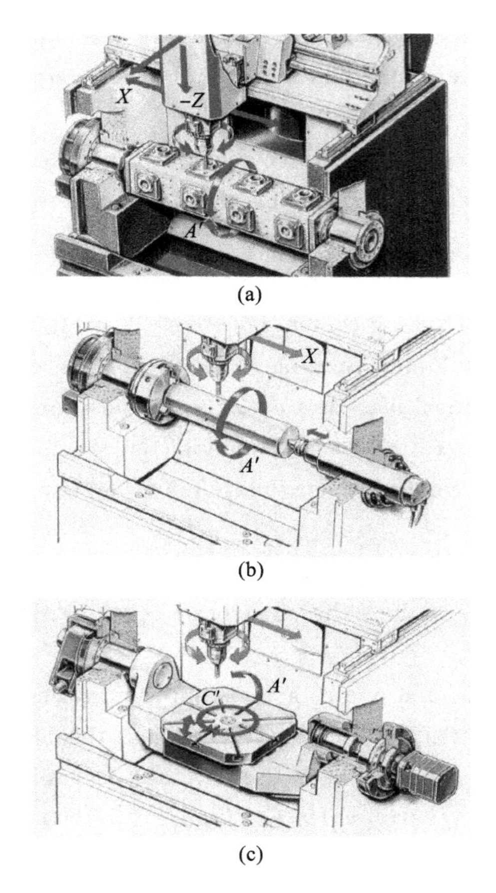

(a)

(b)

(c)

Figure 3-5 Several Expansion Levels of a Three-axis Machine

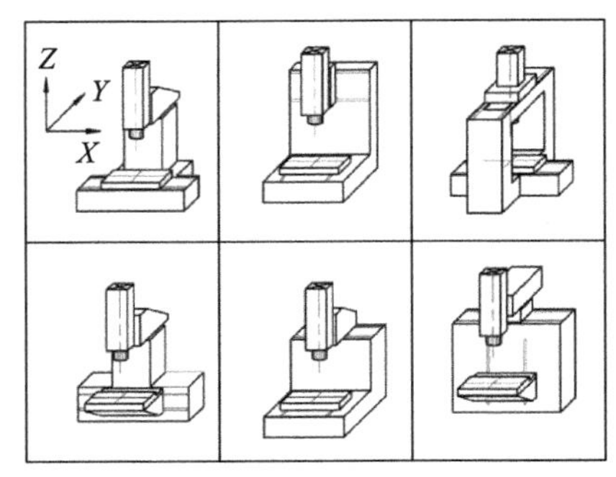

(a)One axis in the workpiece, two axes in the tool

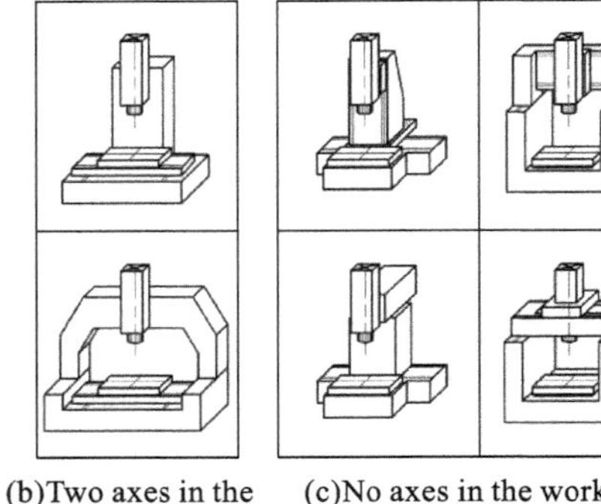

(b)Two axes in the workpiece, one axis in the tool

(c)No axes in the workpiece, three axes in the tool

Figure 3-6 Various Designs of Vertical Three-axis Machining Centers

possible to machine cubical workpieces on four sides in a single setup. If a horizontally/vertically swiveling tool head is used, it is also possible to machine the fifth side. Figure 3-7 shows various types of four-axis machines in a horizontal configuration.

3. Five-Axis Machining Centers

The market share of these machines has increased greatly compared with other machines, and they are now used in both series production and interlinked systems in the automotive industry. Machining centers (Figure 3-8) with five numerically controlled axes can position the tool engagement to any desired point on the workpiece and move along the surface while maintaining the desired angle to the workpiece surface. This universal relative motion between the tool and the workpiece basically can be achieved in three ways below (Figure 3-9).

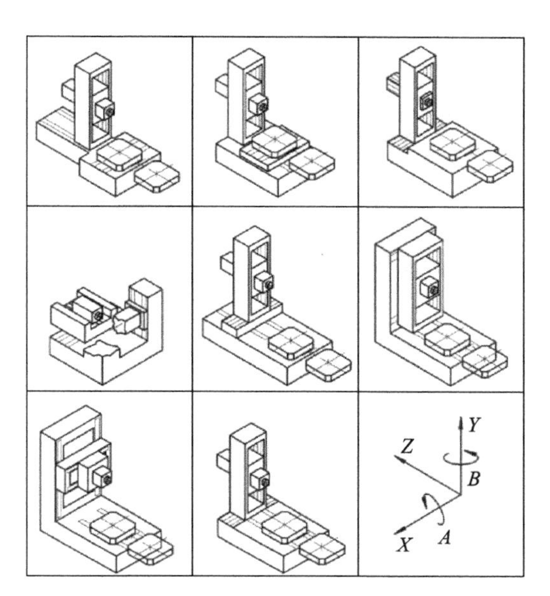

Figure 3-7 Horizontal Four-axis Machining Centers with Pallet Changers

Figure 3-8 Modern Machining Centers with Linear Motors

(1) Using a stationary workpiece and a tool with two swivel[①] axes (Figure 3-9 (a) and Figure 3-9 (b)).

(2) Using a stationary tool axis and a workpiece with a double swivel motion, for example, via a swiveling rotary table (Figure 3-9 (c)).

(3) Using a tool axis and a workpiece that each has a swivel motion, offset by

① swivel ['swɪvl] *n.* 旋转

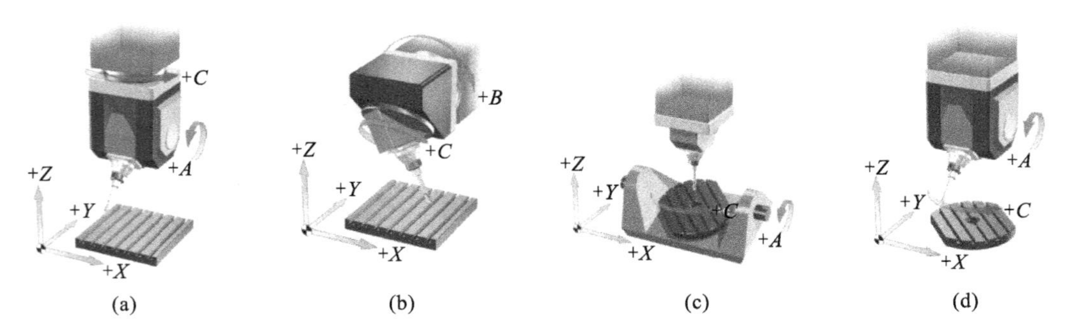

Figure 3-9 Four Options for the Kinematics of Five-axis Machining Centers for 3D Machining

90° relative to each other (Figure 3-9 (d)).

With such machines it is possible not only to create geometrically complex parts but also to use cutter heads with a higher stock-removal rate in place of end mills① or ball-nose cutters② when machining curved surfaces.

4. Multiple-Spindle Machining Centers

All the machines mentioned above can be designed as two-, three-, or four-spindle machines in order to process a number of identical workpieces at the same time. Two-, three-, and four-spindle machines are used most of all in large-scale series production; this also requires the use of multiple clamping fixtures.

With multiple-spindle machines, all the tools have to possess uniform dimensions. An identical length is achieved either by means of preadjusted tools or individually calibratable spindles. Length compensation for the individual spindles is performed in automatic mode by moving to calibrated load cells.

5. Milling/Turning Machining Centers

As shown in Figure 3-10, with modern milling/turning centers, all machining technologies can be implemented with rotating or fixed cutters.

As a contrast to turning/milling centers, an interesting family of machines has emerged that had its origin in machining centers for milling. The point of departure for the development of such machines was the analysis of the ranges and families of parts that on the one hand involve large production quantities but on the other hand have to be machined in different ways.

These are milling/turning parts with an emphasis on milling, that is, not turning/milling parts. These parts require complex six-side machining. They are often repeated parts or small batches that require a CNC program with high

① end mill 端铣刀

② ball-nose cutter 球头立铣刀

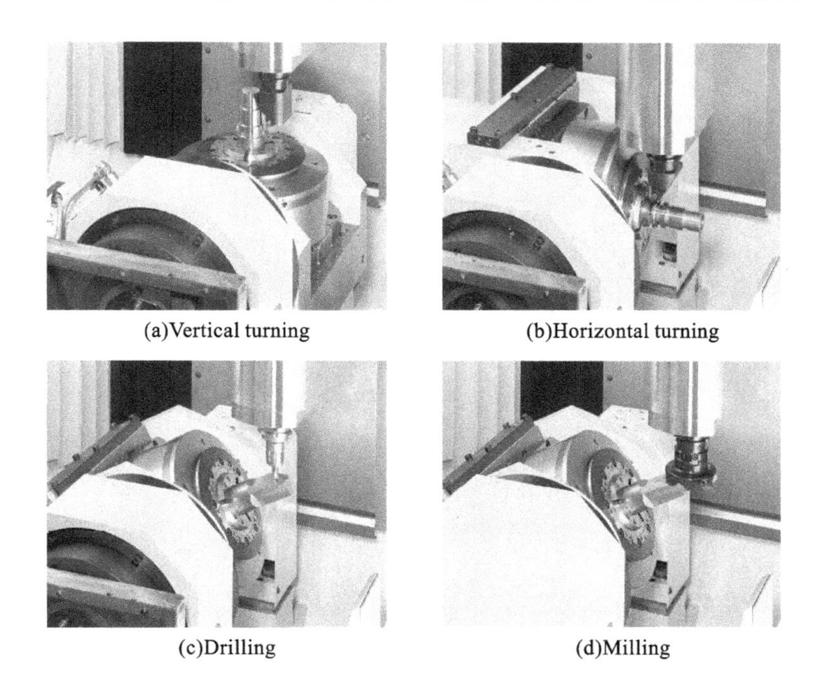

(a)Vertical turning (b)Horizontal turning

(c)Drilling (d)Milling

Figure 3-10 Milling/Turning Machining Centers (Image courtesy of STAMA)

flexibility. The workpieces generally move within a diameter of 60 mm and a length of 100 mm. These centers are also often called bar-machining centers because a wide variety of highly complex workpieces can be produced directly from bar stock. The machining focuses on milling processes, but turning operations can be performed with equal efficiency because the rotary/swivel units are equipped with integrated turning spindles.

The bar stock is fed from the bar magazine① into the main rotary/swivel unit. At this time, the first five sides of the workpiece are machined simultaneously in five axes. To machine the sixth side, the workpiece is transferred to the second rotary/swivel unit. Then the sixth side is machined.

To ensure efficient use of the machine, a tool magazine with about 100 tools is recommended.

Figure 3-11 is a milling/turning center implemented with a counter spindle for machining of the sixth side. The counter spindle (rotational) likewise can be swiveled by 120° (−30° to +90°) and can be moved in the longitudinal direction (X axis).

---

① magazine [ˌmægəˈziːn] *n.* 刀库

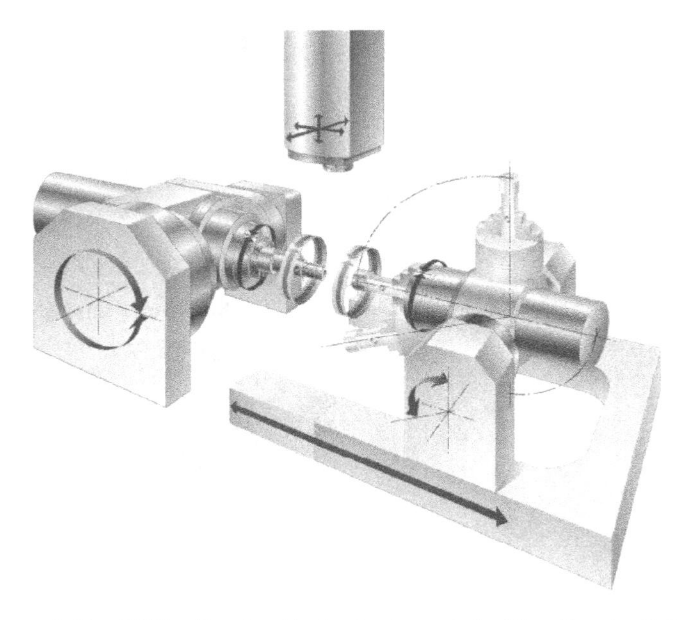

Figure 3-11　Milling/turning Center Implemented with a Counter Spindle for Machining of the Sixth Side (Image courtesy of STAMA)

EXERCISE 4: Word Parts Exercise

Directions: Clues to the meaning of many mechanical words may be found by looking at the word parts. In Part 1, the meaning of a number of prefixes, roots, and suffixes are given. Review Part 1 and then for each word listed in Part 2, do the following things.

(1) Find clues to the meaning of the word by combining the appropriate word parts.

(2) Write your literal translation next to the word, in the middle column.

(3) Look up the actual definition of the word in a dictionary.

(4) Write down the dictionary definition next to the word, in the right hand column.

(5) Compare the two definitions of the word—literal and actual—to see how different the meanings are.

Find some mechanical words in TOPIC 4 and add to the form to help you understand the new words.

Part 1

Prefixes		Roots		Suffixes	
electro	=electric	chron	=time	-ic	=characterized by, like
centro, centri, centr-	=around, center	dynam	=power	-ire	=go
trans-	=across				
hydro-	=water				
syn-	=together				
con-	=with, together				
dia-	=through				

Part 2

Technical Term	Literal Translation	Actual Definition
electromechanical (electro+mechanical)		
synchronize (syn + chron)		
hydrodynamics (hydro+dynam)		
concentric (con + cent + ric)		
ingenious (ingen + ious)		
altimetry (alti + metry)		

3.1.5 Oral Vocabulary

Not all exercises need to be written. Much of your vocabulary instruction can occur during the discussion of reading assignments. Often, through discussion, students can be led to organize the knowledge that they already have about words in order to discover the meaning of a new word.

For instance, the term *traction* is found in a number of different occupational programs. Do you see a smaller, familiar root word within the new word? —In this case, *tract*. Next, please identify other words that have *tract* as a root or base, and you can list those words. For example:

tractor	distraction	attract
contract	detract	contractor
subtract	extract	protractor

Then, please analyze the meanings of these words. What does a tractor do? It pulls things. *Tract* + *or* = pulls + something or someone. The other words can be analyzed in the same manner.

at + *tract* = pulls toward

de + *tract* = pulls away

ex + *tract* = pulls out

con + *tract* = pulls together or with

To get back to the key word, *traction* can then be translated as "the act of pulling". From this brief discussion, you will not only know a "new" word, but also will see the relationships that you may never have seen before between new words and other words you already knew.

By helping to form a pattern of past knowledge, a place—a context for new knowledge can be created. Whenever you see another new word that has the root *tract* in it, you can immediately associate it with all the others you know. As a result, you probably will be able to figure out its meaning.

TOPIC 5: Mechanical Analysis

Modeling and Simulating Mechanical Systems on a Transforming Dicycle

By Danaan Metge, BPG Motors

The Uno Ⅲ is unlike any other vehicles in the world. Originally prototyped as a self-balancing, electric dicycle, the Uno Ⅲ can transform into a motorcycle by deploying[①] a third wheel on the fly, as shown in Figure 3-12. During the transformation to motorcycle mode, the third wheel, stored in Uno mode, is moved forward while the rear traction wheels shift backwards, providing greater stability at high speeds.

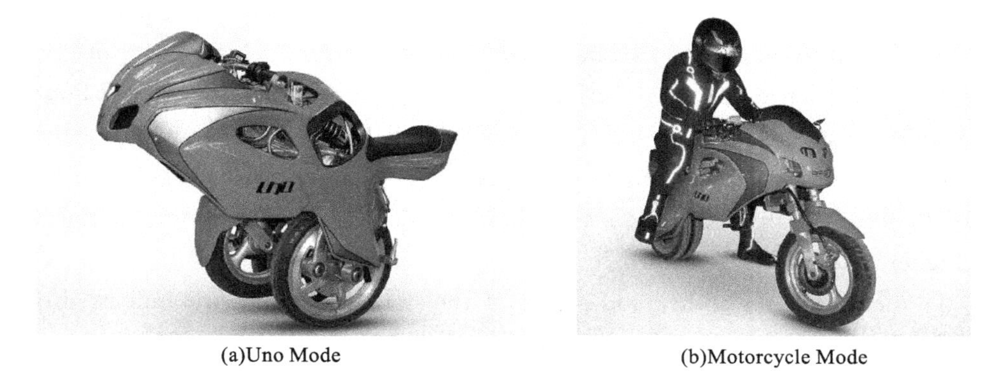

(a)Uno Mode (b)Motorcycle Mode

Figure 3-12 The Uno Ⅲ Dicycle in Uno Mode and in Motorcycle Mode

Much like aircraft control systems, the Uno's controllers must manage roll, pitch, and yaw[②], as well as forward motion and the transformation from dicycle to motorcycle. They must also handle throttle and steering inputs from the rider, as well as side-to-side and front-to-rear shifts in the rider's position. With five independent motors, six gyroscopes and accelerometers, and four potentiometers onboard, testing and tuning the Uno's control systems is a complex task, made even more challenging by the need to ensure rider safety at all times.

When we began the Uno Ⅲ redesign, it was clear that manually tuning controllers and testing on the actual vehicle would be inefficient and risky. Instead, we used Simulink, Simscape, and SimMechanics to model and simulate the Uno's mechanical systems.

During real-world tests, things move so fast, so it is impossible to understand everything that is happening. In simulations, however, we can use Simulink to freeze time and inspect every aspect of the model to get a clear picture of how the mechanics are behaving. We can then see exactly how to improve the control system.

---

① deploy [dɪ'plɔɪ] vt. 配置

② yaw [jɔː] n. （火箭、飞机、宇宙飞船等）偏航

A Brief History of the Uno

Company founder Ben Gulak conceived the idea for the Uno while still a teenager. On a trip to China in 2006, struck by the pollution caused by heavy city traffic, he decided to create an eco-friendly commuter[①] vehicle capable of being driven and stored in congested areas.

Gulak built the original Uno on an angle-iron frame from wheelchair motors, batteries, and gyroscopes[②]. The design won a Grand Award at the Intel International Science and Engineering Fair, and appeared on the cover of Popular Mechanics magazine.

After securing startup funding, Gulak and a team of engineers designed the Uno Ⅱ, which enabled the relatively low-speed dicycle to convert into a motorcycle for higher-speed operation.

To increase the stability and safety of the Uno, the company embarked on a complete redesign for the Uno Ⅲ. This redesign included improvements to the transforming technology and the gyroscopic tilt system.

Modeling the Tilt System

Like a motorcycle, the Uno tilts when turning. The control systems must maintain proper balance throughout each turn. We initially tried to develop and test the balance and tilt control loops independently, but early hardware tests showed us that they were coupled. To better understand these systems and their interdependencies, we ran simulations in Simulink. We could then examine moments of inertia, velocity, control signals, and other system characteristics that would be difficult or impossible to measure in real-world tests.

We built the basic framework by importing the mechanical design from SolidWorks 3D CAD software into SimMechanics (Figure 3-13). To this framework we added the electric motor, a potentiometer, updated ball joints to link the push rods, and the mass of the rider. Via simulation, we applied motor torque and measured the tilt of the system for a range of motor angles.

After postprocessing the simulation results in MATLAB, we developed a transfer function between motor rotation and bike tilt. This analysis enabled us to better understand how the Uno would respond to changes in tilt motor torque (Figure 3-14). We used the same approach to understand the relationship between differential torque applied to the wheels and the radius of the resulting turn.

---

① commuter [kəˈmjuːtə(r)] *n.* 通勤者，经常乘公共车辆往返者

② gyroscope [ˈdʒaɪrəskəʊp] *n.* 陀螺仪

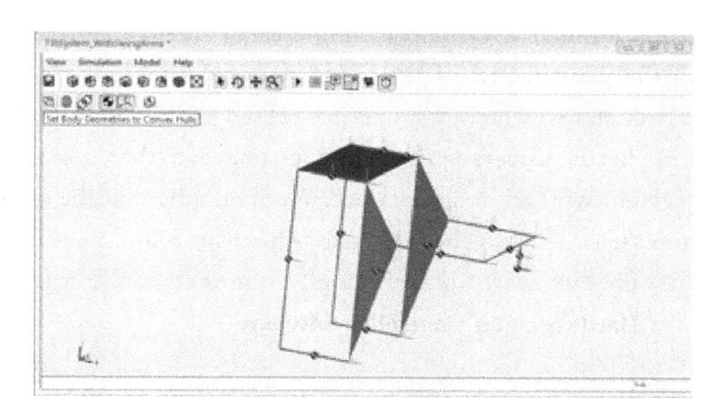

Figure 3-13 SimMechanics Model of the Tilt System

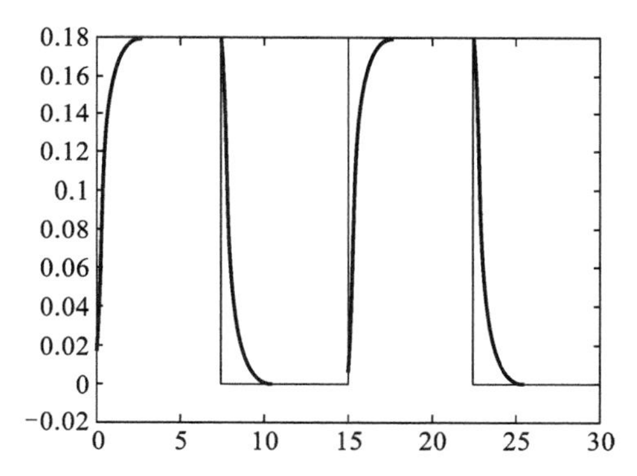

Figure 3-14 Graph of the Uno Tilt System Step Response to a Tilt Motor Command (Rectangular wave = the ideal response; Oval wave = the simulated response)

The complete tilt model enabled us to experiment with different sensors. We supplemented the Inertial Measurement Unit with a potentiometer① in the tilt system. Based on simulation results acquired from the Simulink model, we later moved to a higher-resolution analog-to-digital converter, which we ultimately used in the final design.

Tuning Controller Gains

BPG software engineers developed the proportional integral derivative (PID) controllers for the Uno Ⅲ in C, intending to use data gathered during hardware tests to tune the controller gains. In practice, this proved to be impractical using a traditional Ziegler-Nichols method because when we increased the proportional gain, the system's output never stabilized as we expected.

① potentiometer [pəˌtenʃiˈɒmɪtə] *n.* 电位计

To resolve this problem, we built a simple PID controller in Simulink and ran simulations with the plant model of the tilt system. We placed scopes throughout the model to collect data, which we post-processed in MATLAB. This analysis enabled us to first better understand the tilt system and then tune the controller gains until the system worked in simulation. We then adjusted the controller gains in our C code to match the ratio that we had verified in Simulink. The adjustment led to a breakthrough—we were able to actually ride the Uno Ⅲ for the first time.

Modifying the Hardware and Controller Software

In addition to helping us tune controller gains, Simulink simulations also provided design insights that led to control algorithm changes and hardware modifications. For example, we ran multiple simulations in Simulink to see if gain scheduling could improve the tilt system. We found that a stepwise① loop that used one set of gains for tilt angles between $-3°$ and $+3°$ and different gain values for progressively larger tilt angle ranges produced better overall performance than a linear PID loop.

Later, we used Simulink and SimMechanics to explore mechanical changes of the system. In one instance, we ran simulations to ensure that the tilt motor had enough torque to move the 350-pound② Uno and its rider from side to side. After conducting these simulations, we weren't convinced that the motor at its current size would be capable of moving the Uno quickly enough. While the results were not definitive, we decided to err on the side of caution and use a larger motor.

Simulink simulations also helped us identify a deficiency with our analog-to-digital converter (ADC). Using some basic ADC blocks in Simulink, we built and simulated a simple model that helped us identify dead spots in our control algorithm that were affecting performance. To address the problem, we replaced the hardware ADC on the Uno with one that had a resolution four times higher.

Simulating the Transition and Power Systems

Because the Uno's transformation from dicycle to motorcycle occurs while the vehicle is moving, ensuring the safety of this transition is a top priority. To simulate the transition with Simulink and SimMechanics, we combined a simple model of the balance control system with an inverted pendulum model. We then estimated the position of the Uno's center of mass relative to the traction wheels at several states in the transition process, and used Simulink to verify that the

---

① stepwise['stepwaɪz] *adj*. 逐步的，步进的

② 1 pound=0.453 kg

mechanical model was controllable for each state.

As the Uno moved from prototype to preproduction, we were expanding our use of Simulink to model and simulate aspects of the Uno that would be too costly, dangerous, or time-consuming to experiment with on the actual hardware. We recently used SimPowerSystems to model the Uno's 48V power system, including the batteries, switches, and motors, to capture inductance and capacitance effects. Instead of hooking up probes and meters to the actual power system, we simulated it to track down the source of the spikes that we had identified during testing.

Going forward, we plan to reuse the model for additional reliability checks and to estimate battery life for various drive cycles and conditions.

EXERCISE 5: Word Search and Word Puzzle

Directions: Vocabulary exercises involving word searches, word puzzles and games can also be used to learn new words and their meanings. In addition, they can help you review the meanings of words you have already learned. For example, now, circle all the words you find in the word search box. The letters can be horizontal, vertical, or sometimes diagonal.

Sensor mechanics dicycle rear motor controller

System code ratio algorithm simulation

A	D	W	N	R	E	N	Y	N	T	M	W	S
E	I	A	M	V	T	U	T	C	R	R	A	I
T	C	E	C	E	B	L	R	O	A	I	E	M
Q	Y	Q	A	Q	T	I	F	D	T	E	L	U
E	C	L	B	R	E	S	P	E	I	N	J	L
P	L	X	V	E	N	O	Y	T	O	J	E	A
M	E	C	H	A	N	I	C	S	E	H	Y	T
C	O	N	T	R	O	L	L	E	R	I	I	I
U	W	T	A	L	G	O	R	I	T	H	M	O
T	E	Z	O	E	A	D	W	Q	P	U	Q	N
B	E	T	W	R	O	S	N	E	S	C	U	O

3.2 Reading Experience Ⅱ: Developing Comprehension Skills

Understanding what you read while you are reading it—that's comprehension. Reading is the process of translating symbols into ideas. It sounds simple, but it is a complex process. There are a number of abilities involved in reading, including those that follow.

Ability to understand graphic symbols. This means students must be able to make sense out of the little black letters printed on a page. In addition, they must be able to understand the meaning of lines, numerals, and drawings on a chart or diagram. Vocational-technical subjects are loaded with symbols that students must understand in order to complete a task.

Ability to react to sense images (sight, sound, taste, touch, and smell) suggested by words. While reading a description of a plant, for example, a horticulture student should be able to form a mental picture of the healthy plant, the color and the smell of its blossoms, the texture of its leaves, and the taste of its fruit.

Ability to perceive relationships. you must be able to distinguish between cause and effect, general and specific, whole and part, smaller and larger. In many occupational areas, it is particularly important to understand time and sequence (e. g. what must happen first, second, and third in a critical nursing procedure).

Ability to follow directions. Vocational-technical students must be able to read a series of instructions and then take a course of action. Sometimes the directions are complex, and most time there is little room for error. Therefore, words must not be misunderstood, and parts cannot be overlooked. The skill of following directions does not come naturally, it must be learned.

Ability to understand written units of increasing size. You need to understand not only the meanings of individual words, but how they are used in phrases, sentences, and paragraphs. As the written unit gets larger, the difficulties of comprehension tend to increase. You may read a whole paragraph or book section, for example, to get the most out of it.

Ability to make inferences and draw conclusions. This means that you must go beyond the facts presented in the reading and try to anticipate the results or effects that might follow. It is a simple matter, for example, for a retail you are going to read that the price of copper has gone up dramatically. It is much more difficult for you to determine what that means in terms of the sale of electrical appliances or the design of plumbing supplies.

3.2.1 Levels of Comprehension

There are three levels of comprehension: literal, interpretive, and applied. *Literal* means word-for-word. At the **literal level** of comprehension, therefore, one is looking for the exact meanings of the words. To do this, students need to be able to identify the main ideas, spot relevant details, observe sequence, follow directions, and note conclusions.

To *interpret* means to go beyond what the text actually says to what it really means. Thus, at the **interpretive level** of comprehension, what determines the author's purpose, notices, causes and effects may not have been directly stated, so one needs to make inferences and draw conclusions, and so forth. In other words, students need to be able to read between the lines and fill in the gaps left by the author.

At the **applied level**, one uses the information the author has provided to do something—rebuild an engine, construct a porch, give a blood test. Since the ability "to do"—to perform occupational skills—is a primary goal of all vocational-technical programs, developing reading comprehension skills to this level is an absolute necessity.

3.2.2 Reading for Main Ideas and Details

The most basic skill in reading comprehension is the ability to identify the main ideas and important details. This skill is founded upon accurate comprehension of technical terms and phrases, which is why vocabulary development is essential. Without the ability to find main ideas and details, you cannot hope to figure out the author's meaning or recall information in order to apply it to a job or task.

Read the following reading—TOPIC 6: The Role of Mechanical Engineer, then finish EXERCISE 6.

TOPIC 6: The Role of Mechanical Engineer

Who are Mechanical Engineers?

The field of mechanical engineering encompasses the properties of forces, materials, energy, fluids, and motion and the application of those elements to devise[1] products that advance society and improve people's lives. The U. S. Department of Labor describes the profession as follows.

Mechanical engineers research, develop, design, manufacture and test tools, engines, machines, and other mechanical devices. They work on power-producing machines such as electricity-producing generators, internal-combustion[2] engines, steam and gas turbines[3], and jet and rocket engines. They also develop power-using machines such as refrigeration and air-conditioning equipment, robots used in manufacturing, machine tools, materials handling systems, and industrial production equipment.

Mechanical engineers are known for their broad scope of expertise and for working on a wide range of machines. They design machinery and power-transmission equipment using various types of gears as building-block components. Just a few examples include the micro electromechanical acceleration sensors used in automobile air bags; heating, ventilation[4], and air-conditioning systems in office buildings; heavy off-road[5] construction equipment; hybrid[6] gas-electric vehicles; gears, hearings, and other machine components (Figure 3-15); artificial hip implants[7]; deep-sea research vessels[8]; robotic manufacturing systems; replacement heart valves; noninvasive[9] equipment for defecting explosives[10]; and planetary exploration spacecraft, as shown in Figure 3-16, the Mars Exploration Rover is a mobile geology laboratory used to study the history of water on Mars. Mechanical engineers contributed to the design, propulsion[11], thermal control, and other aspects of these vehicles.

Based on employment statistics, mechanical engineering is the third-largest discipline among the five traditional engineering fields, and it is often described as offering the greatest flexibility of career choices. In 2002, approximately 215 000

---

① devise [dɪ'vaɪz] *vt.* 设计；想出；发明；图谋；遗赠给

② combustion [kəm'bʌstʃən] *n.* 燃烧，氧化；骚动

③ turbine ['tɜːbaɪn] *n.* 涡轮；涡轮机

④ ventilation [ˌventɪ'leɪʃn] *n.* 通风设备；空气流通

⑤ off-road [ɒfrəʊd] *adj.* 越野的

⑥ hybrid ['haɪbrɪd] *n.* 杂种，混血儿；混合物 *adj.* 混合的；杂种的

⑦ artificial hip implant 人工髋关节植入物

⑧ vessels['veslz] *n.* (vessel 的复数)血管；船舶；容器

⑨ noninvasive [ˌnɒnɪn'veɪsɪv] *adj.* 非侵略性的；无攻击性的

⑩ explosive [ɪk'spləʊsɪv] *n.* 炸药，爆炸物；[*pl.*]爆破器材

⑪ propulsion [prə'pʌlʃn] *n.* 推进；推进力

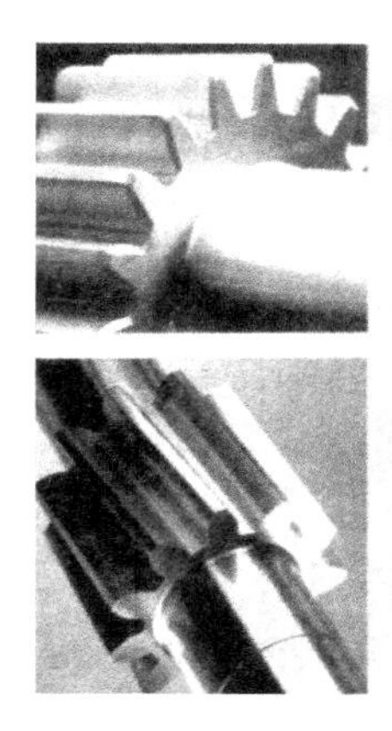

Figure 3-15 Machine Components

Reprinted with permission of Niagara Gear Corporation, Boston Gear Corporation, and W. M. Berg Incorporated.

people were employed as mechanical engineers in the United States, a population representing 15% of all engineers. The discipline is closely related to the technical areas of industrial (194 000 people), aerospace (78 000) and nuclear (16 000) engineering, since each of those fields evolved historically as a spin-off from mechanical engineering. Together, the fields of mechanical, industrial, aerospace, and nuclear engineering account for about 34% of all engineers. Other specializations that are encountered frequently in the mechanical engineering profession include automotive and manufacturing engineering. While mechanical engineering often is regarded as the broadest of the traditional engineering fields, there are many opportunities for specialization in a certain industry or technology that interests you. For example, an engineer in the aviation① industry might focus his/her career on advanced technologies for cooling turbine blades in jet engines or fly-by-wire systems for controlling an aircraft's flight.

Above all else, mechanical engineers make hardware that works. An engineer's contribution to a company or another organization ultimately is evaluated based on whether the product functions as it should. Mechanical engineers design equipment, it is produced by companies, and it is then sold to the public or to industrial customers. In the process of that business cycle, some aspects of the customer's life are improved, and society as a whole benefits from the technical advances and additional opportunities are offered by engineering research and development.

---

① aviation [ˌeɪviˈeɪʃn] *n.* 航空；飞行术；飞机制造业

Figure 3-16 Planetary Exploration Spacecraft

Reprinted with permission of NASA.

Mechanical engineering isn't all about numbers, calculations, computers, gears and grease[①]. At its heart, the profession is driven by the desire to advance society through technology. The American Society of Mechanical Engineers (ASME) surveyed its members in order to identify the major accomplishments of mechanical engineers. This professional society is the primary organization that represents and serves the mechanical engineering community in the United States and internationally. This "top-ten" list of achievements, summarized in Table 3-3, will be useful for you to understand better whom mechanical engineers are and to appreciate the contributions that mechanical engineers have made to your world. In descending order of the accomplishment's perceived impact on society, the following milestones were recognized in the survey.

Table 3-3 The Top-ten Achievements of the Mechanical Engineering Profession

1. The automobile	6. Integrated-circuit mass production
2. The Apollo program	7. Air conditioning and refrigeration
3. Power generation	8. Computer-aided engineering technology
4. Agricultural mechanization	9. Bioengineering
5. The airplane	10. Codes and standards

Compiled by the American Society of Mechanical Engineers.

The following is the description of some achievement in Table 3-3.

① grease [gri:s] *n.* 油膏，油脂

(1) **The automobile.** The development and commercialization of the automobile were judged as the profession's most significant achievement in the twentieth century. Two factors responsible for the growth of automotive technology were high-power, lightweight engines and efficient processes for mass manufacturing. German engineer Nicolaus Otto is credited with designing the first practical four-stroke internal-combustion engine. After untold effort by engineers, it is the power source of choice for most automobiles today. Figure 3-17 is an example of eight-cylinder 5.7 liter engine. In addition to engine improvements, competition in the automobile market has lead to advances in the areas of safety, fuel economy, comfort, and emission control.

Figure 3-17 An Example of Eight-cylinder 5.7 Liter Engine

2002 General Motors Corporation. Used with permission of GM Media Archives.

The ASME recognized not only the automobile's invention but also the manufacturing technologies behind it. Through the latter, millions of vehicles have been produced inexpensively enough that the average family can afford one. Quite aside from his efforts of designing vehicles, Henry Ford pioneered the techniques of assembly-line mass production that enabled consumers from across the economic spectrum① to purchase and own automobiles. Having spawned② jobs in the machine tools, raw materials, and service industries, the automobile has grown to become a key component of the world's economy. From minivans③ to stock cars racing to Saturday night cruising, the automobile—one of the key contributions of

① spectrum ['spektrəm] *n.* 范围，领域，系列

② spawn[spɔːn] *vi.* 大量生产

③ minivan['mɪnivæn] *n.* 小型载客车，小型面包车

mechanical engineering—has had an ubiquitous[①] influence on our society and culture.

(2) **The airplane.** The development of the airplane and related technologies for safe powered flight were also recognized by the ASME as key achievements of the profession. Commercial passenger aviation has created travel opportunities for business and recreational purposes, and international travel in particular has made the world become a smaller and more interconnected place.

Mechanical engineers have developed or contributed to nearly every aspect of aviation technology. One of the main contributions has been in the area of propulsion[②]. Early airplanes were powered by piston[③]-driven internal-combustion engines, such as the 12-horsepower engine that was used in the first Wright Flyer. By contrast, the General Electric Corporation's engines that power some Boeing 777 jetliners can develop a maximum thrust of over 100 000 pounds. Advancements in high-performance military aircraft include vectored turbofan engines that enable the pilot to redirect the engine's thrust for vertical takeoffs and landings. Mechanical engineers design the combustion system turbines, and control systems of such advanced jet engines. They have also spearheaded[④] the discovery and evolution of lightweight aerospace-grade materials, including titanium alloys and graphite fiber-reinforced epoxy[⑤] composites. As shown in Figure 3-18, a scale model of an aircraft is being prepared for tests in a subsonic wind tunnel.

(3) **Computer-aided engineering technology.** The term "computer-aided engineering" (CAE) refers to a wide range of automation technologies in mechanical engineering, and it encompasses the use of computers for performing calculations, preparing technical drawings, simulating performance, and controlling machine tools in a factory, as shown in Figure 3-19, mechanical engineers use computers to analyze the airflow around the space shuttle orbiter[⑥] during flight. Mechanical engineers don't design the architecture of a computer, but they do use computers on a day-to-day basis. Over the past several decades, computing and information technologies have changed the manner in which mechanical engineering is practiced. Most mechanical engineers have access to advanced computer-aided design and

---

① ubiquitous [juːˈbɪkwɪtəs] *adj.* 无所不在的；普遍存在的

② propulsion [prəˈpʌlʃn] *n.* 推进(力)；推进器

③ piston [ˈpɪstən] *n.* 活塞

④ spearhead [ˈspɪəhed] *n.* 先锋；前锋；先头部队

⑤ epoxy [ɪˈpɒksi] *adj.* 环氧树脂的

⑥ orbiter [ˈɔːbɪtə] *n.* (尤指在轨道上运行的)轨道飞行器

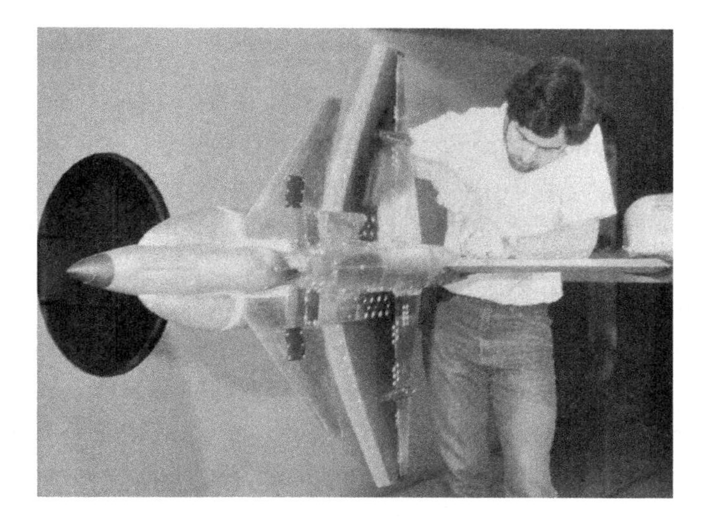

Figure 3-18 Preparation for Tests of an Aircraft

Reprinted with permission of NASA.

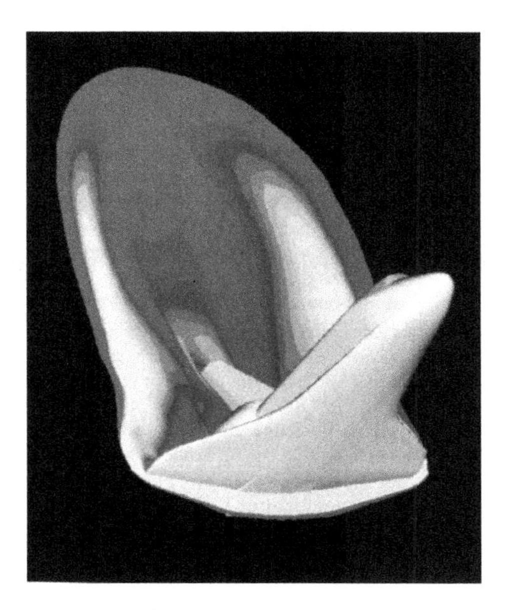

Figure 3-19 Analysis of the Airflow around the Space Shuttle Orbiter with Computers

Reprinted with permission of NASA.

analysis software, information databases, and computer-controlled prototyping equipment. In some industries, these CAE technologies have replaced traditional paper-based design and analysis methods.

Mechanical design automation began in earnest during the late 1950s, primarily for aerospace and defense-related systems. At that time, computer hardware, graphics technologies, and programming were at relatively low levels when compared to today's capabilities. Because of the expense involved, most early applications concentrated on computer drafting. However, the costs involved in using state-of-the-art CAE hardware and software have crept downward sufficiently so that even small-sized companies now develop products by using integrated computer-aided design and manufacturing systems.

In large multinational corporations, design teams and technical information are distributed around the world, and computer networks are used to design products 24 hours a day. As an example, the Boeing 777 was the first commercial airliner to be developed through a paperless computer-aided design process. The 777's design began in the early 1990s, and a new computer infrastructure had to be created specifically for the design engineers. Conventional paper-and-pencil drafting services were nearly eliminated. Computer-aided design analysis, and manufacturing activities were integrated across some 200 design teams that were spread over 17 time zones. Through the extensive usage of CAE tools, designers were able to check part-to-part fits in a virtual, simulated environment before any hardware was even produced. By constructing and testing fewer physical mock-ups① and prototypes, the aircraft was brought to market quicker and more economically than would have otherwise been possible.

EXERCISE 6: Reading for Main Ideas and Details

(1) What makes the mechanical engineer's career choice the most flexible?

(2) Under each of the two categories below, write the great achievements of mechanical engineers.

---

① mock-up 实体模型，实尺度模型

	Fields	Great achievements of mechanical engineers
1	The automobile	
2	The airplane	
3	Computer-aided engineering technology	

(3) Read the sentences: "mechanical engineers make hardware that works. An engineer's contribution to a company or another organization ultimately is evaluated based on whether the product functions as it should. Mechanical engineers design equipment, it is produced by companies, and it is then sold to the public or to industrial customers. In the process of that business cycle, some aspects of the customer's life are improved, and society as a whole benefits from the technical advances and additional opportunities are offered by engineering research and development."

- Is this sentence related to the reading selection in any way?

• Do you think this sentence summarizes what the author has implied in the reading?

• What then, would you say is the main idea of the reading selection?

3.2.3 Reading for Organization

Most information in textbooks and learning guides is (or should be) presented using basic patterns that stress a relationship among the ideas. In order to gain the maximum amount of meanings from your reading, you must be able to recognize the patterns and understand the relationships. The following are descriptions of the basic patterns.

Time sequence. In this pattern, events or procedures are organized in the sequence (or order) in which they take place. It might be a historical sequence. For example:

The vacuum tube was invented in the early part of the century, transistors were developed later, and recently microprocessor chips have come into use.

This type of pattern is often employed in describing a technical process that takes place over time. For example:

The heated steel will first have a dull cherry-red color. Then, as heat continues to be applied, it will begin to turn bright red, then orange, and then yellow. Finally, it will turn white and sparks will begin to fly.

Comparison contrast. In this pattern, the author may show similarities and differences between concepts with which the reader is already familiar and a new concept. For example:

A mango has the orange-colored flesh of a peach but is much stronger and spicier in flavor.

Another device used is advantages and disadvantages. And differences in characteristics or purposes can also be used. For example:

A monkey wrench is a convenient all-purpose tool, whereas an open-end wrench is used where strength and good fit are important.

Cause and effect. This pattern of organization is often used in technical literature that deals with troubleshooting or with adjusting mechanisms. For example:

Turn the adjustment knob to the right [cause] to increase the height of the flame [effect].

A loud hum[effect] may be the result of a poor ground in the circuit [cause].

Step-by-step procedure. One of the most prevalent organizational patterns used in vocational-technical materials is that of describing technical procedures in a step-by-step sequence. A great many procedures lend themselves to this pattern. However, the written steps are not always clearly labeled *Step* 1, *Step* 2, *Step* 3, and so on. You may need help in discerning step-by-step instruction, especially when it is obscured in a narrative form. For example:

It is important to hold the pilot light button down for one minute before lighting the main burner.

Simple listing. Lists of facts, characteristics, safety precautions, do's and don'ts, and other items are very common in vocational-technical materials. The items on a list may be presented in some specific order (e. g. importance, time), or they may be presented in random order.

Categorization. Grouping things by some shared characteristics is used particularly in descriptive information. For example, in describing woods the categories might be as follows: type of grain (open-grained/close-grained), hardness, texture, color, and so on. Categorization can help the reader deal more effectively with a large number of facts.

TOPIC 7: Manufacturing Managements

Manufacturing Engineering Budgets

The facilities program description touches on many aspects classified as budget matters and project management matters.

The budget represents the funds allocated for a specific purpose, in the case for the purchase of capital equipment. It also shows by its size and adequacy the priority level given by senior management to this aspect of the business. A budget that is set low by comparison to previous budgets indicates that management is not interested in business expansion, but will be satisfied with maintaining the company's current position vis-à-vis[①] market penetration and profits. A budget that is set high indicates that management wants to take market share away from competitors and place the company in position to become more profitable. Therefore, by evaluating the size of the budget as compared previous years, manufacturing engineering can easily gauge the intent of senior management and act accordingly. This is always true and provides guidelines for how manufacturing engineering should act once the budget is presented.

However, this is only half the story. Manufacturing engineering has considerable opportunity to influence management in deciding whether to hold the line or to set an aggressive budget. As the technical arm of manufacturing, it must make sure that senior management understands what funds are required for the company to maintain its present capability or capacity, and what would be required in addition to improve capability or capacity. Therefore, it must translate matters from a technical sense to a financial sense, that is, the cost versus opportunity relationship, keeping in mind that technical improvements must result in improved profit-making opportunities to be of use to an industrial enterprise. The algorithm the manager of manufacturing engineering must keep in mind is that technical improvement equals opportunity for profit minus costs:

$$T = \mathrm{OP} - C$$

and that opportunity for profit minus costs must be greater than zero:

$$\mathrm{OP} - C > 0$$

With these principles in mind, manufacturing engineering can proceed to put together proposed budgets, for both the current period and the forecast period. Even such items as upkeep of plant and facility—for example, painting offices—must satisfy the algorithm; that is, walls should not be painted unless the benefits outweigh the costs. Therefore manufacturing engineering must go beyond the reasoning that "it would be a nice thing to do" to justify a proposed budget item. In the case of painting offices, a justification that can be related to improved

① vis-à-vis *prep.* 对于，关于；和……相比

opportunity for profit must be shown. A reason to paint a sales office—to influence potential customers favorably—would not be sufficient for a manufacturing floor dispatch[①] office. Perhaps the dispatch office should be painted to alleviate a safety problem that, if not solved, could cost the company funds in lost production, fines, and so forth. These mundane examples point out that manufacturing engineering must be aware that every dollar they propose to be included in the capital equipment budget must relate directly to improved profit opportunity.

When manufacturing engineering goes through the objectives and goals process, it is automatically going through a capital budget proposal process. Each project finally approved by the objectives and goals system satisfies the algorithms presented above. Whether a specific project reaches the approved budget depends largely on the degree of positiveness with which it satisfies the expression $OP-C>0$. In the budget process manufacturing engineering is competing with all other company functions and subfunctions for limited funds. Therefore, it must strive to place only projects with large positive values of $OP-C$ calculations in its submittal of a proposed budget. Of course, it must not lose sight of the fact that projects must be consistent with overall goals. If manufacturing engineering follows the simple procedure of developing capital expenditure projects as part of the objectives and goals procedure, and then assuring that the profitability algorithms are adhered to, it will have substantial success in achieving funding for its programs.

In making a facility program budget it is desirable to classify projects in accordance with the major category to be accomplished by implementing the project. The classifications show what the major thrust of the company is for any given budget period, that is, whether the company is attempting to enlarge its market share or trying to maintain the status quo.

Classifications vary among companies, but as a minimum, for an adequate description of all capital equipment projects, they should include the following categories.

(1) Rebuild—used for projects concerned with taking existing equipment out of service and rebuilding it as new or with slightly enhanced capabilities.

(2) Replace—used for direct replacement projects, replacing one piece of equipment with another, usually of the same type and capability.

(3) Cost improvement—addition of equipment that allows the company to

---

① dispatch [dɪˈspætʃ] *n.* 调度，派遣

produce its products at lower costs, hence improved productivity.

(4) Capacity addition—addition of equipment that allows more products to be produced than before or a new type of products to be produced.

(5) Safety and environment—equipment purchased primarily to meet company, industry, or government regulations①.

(6) Miscellaneous—used for capital equipment projects that do not fall into any other categories.

Classification of all projects helps senior management to visualize where funds will be spent. It also makes the intent of the company visible. If the company states in its business plan that it wants to increase its market share, and all of its capital equipment projects are in the replace and rebuild categories, it is clear that the actual plans and the strategy are not in agreement. These types of mismatches occur frequently, and the check afforded by classifying capital equipment projects brings them into the open and may lead to either a change in funding levels for capital equipment projects or an honest assessment that the strategy cannot be accomplished and should be changed. Classification, then, can be thought of as a check on the system. If the majority of the equipment project classifications agree with the stated strategy as reflected in the objectives and goals, then the company's planning systems are working well and one would tend to have confidence in its ability to achieve the stated goals. If the converse is true, there is a management problem that should be addressed.

The final budget matter that concerns manufacturing engineering is the question of how much to propose. A proposal for spending at a level that is impossible to achieve will be totally discounted, and vital projects may inadvertently be discarded. To maintain its credibility, manufacturing engineering must know what level of expenditure is reasonable and what is unreasonable.

What is reasonable, of course, depends on the situation. Fortunately, there is a way to check whether a budget proposal is in the realm of reasonableness. Assuming that a company intends to stay in business and intends at least to maintain its current market level, it is fair to say that the capital equipment investment budget should be equal to or greater than the depreciation amount. This means that to stay healthy over a 3 to 5-year period, a company should invest as

---

① regulation [ˌregjuˈleɪʃn] *n.* 管理；规则；校准

much money in capital equipment as it depreciates[1] for bookkeeping and tax purposes. If it does less, the company is shrinking in size and its net worth is decreasing.

With this fact in mind the manager of manufacturing engineering can set budget targets. If the company intends to increase its market share for its product line, then the manager would be justified in proposing a budget that is higher than depreciation. I believe that the increment should be a one-for-one increase based on the amount of market penetration increase aimed at. At the other end of the spectrum, the manager of manufacturing engineering should protest if investments are kept below depreciation for the entire long-range forecast period. Such a situation is a defacto slow exiting from the business, and it is the manager's duty to point this out. Therefore the general budget target should be equal to capital equipment depreciation plus or minus amounts corresponding to current year situations, with concern for consecutive negative years. Depreciation is a very simple yardstick for budget proposals. Depreciation accounts are calculable well in advance, since they are based entirely on existing capital equipment and firm depreciation rules. This makes it easy for manufacturing engineering to obtain the figures and set visible targets for budget preparation.

EXERCISE 7: Reading for Organization

Directions: Since good writers want to be understood, they use signal words to let the reader know what pattern they are using. The emphasis in observing paragraph organization, then, is on noting the key words that indicate the patterns being used and seeing relationships among the ideas being presented.

Such words as *first*, *next*, *third*, and *finally* are clues that the author is listing details or steps in a procedure. *By contrast*, *on the other hand*, *however*, and *yet* are signals of comparison/contrast statements. Clues to chronological organization are such words as *initially*, *then*, *soon*, *later*, *after that*, and *at last*.

Paying attention to signal words within paragraphs can greatly aid you in reading longer selections, since whole chapters are often organized in this way.

(1) For the relationships between manufacturing engineering and management (TOPIC 7), fill in the chart using comparison contrast.

---

① depreciate [dɪ'priːʃieɪt] *vt.* 使贬值；贬低

Management intentions		Manufacturing engineering intentions	
Set low budget	Set high budget	Hold the line	Set an aggressive budget
How do they do this?		How could they do this?	

(2) Meanwhile, the entire reading selection is organized by categorization, which is mainly about the budget matter that concerns manufacturing engineering. Can you list out the budget matter that manufacturing engineering needs to care about?

3.3 Reading Experience Ⅲ: Developing Graphics Exercises

Modern textbooks and instructional materials in all vocational-technical programs use a far great variety and quantity of graphic material. Photographs, draw-logs, paintings, cartoons, graphs, charts, diagrams, tables, and maps all appear frequently in today's materials.

The purpose served by graphs of all kinds are to reinforce and clarify concepts contained within the printed text and to make the text's appearance more interesting and attractive. Although most visual presentations have features in common, each kind has its own special way of conveying information. This is true even within a particular class of graphic material.

By their nature, some graphs may be more suited than others to the presentation of a particular piece of information. For example, which would be of more help in showing how an electrical circuit works, a schematic diagram, or a gorgeous photograph of an electrical device that took first place in the science fair? The schematic diagram would, of course, be more helpful.

The following discussion illustrates how oral questioning based on a graph may move through all levels of thinking and comprehension, from low to high.

First, focus on the title of the graph. It is always important to pay attention to the title of a graph, because it establishes the main idea and indicates purpose. This is a point to be kept in mind when producing your own visuals: always give them titles. Although the content and meaning of a graph or chart may be perfectly obvious to its creator, it may remain a mystery to another person unless a hint (such as a title) is given.

Second, learn necessary vocabulary. One way to do this is to ask yourself questions that lead to discover the meanings of key terms.

Third, point out the various parts of the graph and identify the details that support the main idea. Then, you can ask questions about the details. Remember, your purpose is not just to learn content. You are also trying to learn a skill—one that can transfer to reading any graphs, whether they occur in your programs, in other subject areas, on the jobs, or in personal readings.

Fourth, begin asking questions about the factual content of the graph.

Fifth, after getting factual information from the graph, begin asking higher-order questions to help yourself interpret and analyze the graph.

Finally, pursue a similar analysis of other parts of the graph.

3.3.1 Types of Graphs and How to Read Them

Graphs are visual representations of numerical data showing comparisons and relationships. Different types of graphs are shown in Table 3-4.

Table 3-4 Different Types of Graphs

<table>
<tr><th>Types of Graphs</th><th>Description and Purpose</th><th>How to Read Graph</th></tr>
<tr><td>Line 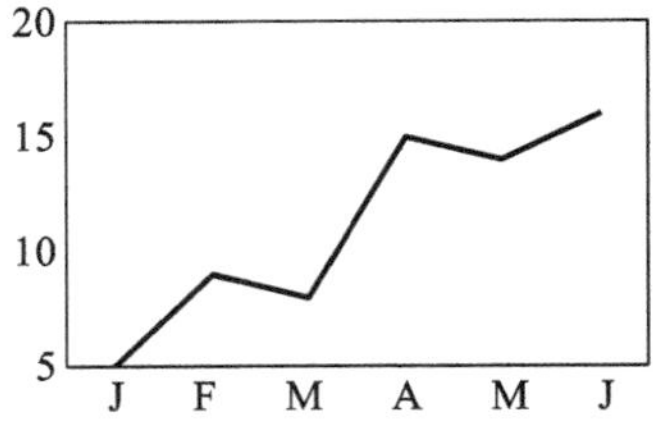</td><td>(1) It indicates precise relationship between two sets of data.
(2) Each point on the graph represents the two variables in relation to each other.
(3) It is the most accurate type of graph and shows development taking place and trends</td><td rowspan="5">(1) Note title and type of graph. These indicate purpose and main idea.
(2) Note representations of data. Read both vertical and horizontal titles to decide what is being compared (e. g. dollars/year, pounds/acre).
(3) Note scale. What are the increments of increase/decrease? Be alert to alternations within the pattern that can change appearance and cause misinterpretations.
(4) Read the key. It indicates the meaning of symbols. Color codes and surface patterns (e. g. cross-hatching, dots) are often used.
(5) Note symbols within the graph. They may be merely decrease or may be meaningful components (consult the key).
(6) Read for the information.
(7) Criticize. Make inferences and draw conclusions based on data. What applications are possible?
(8) Relate to text material</td></tr>
<tr><td>Bar 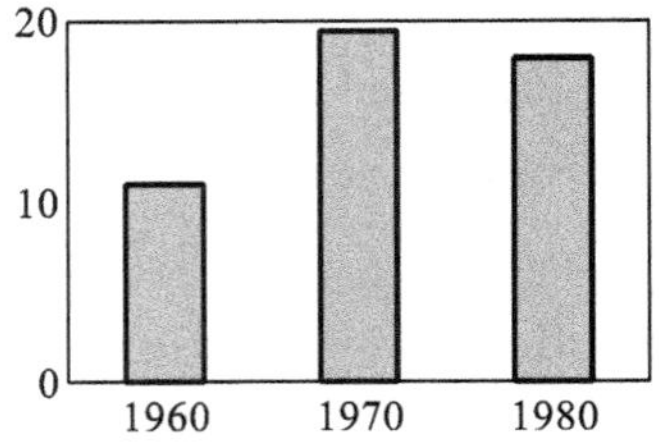</td><td>(1) It permits comparison of a small number of values (fewer than ten) taken at different time or representing different age, groups, countries, sexes, etc.
(2) Presentation may be made vertically or horizontally.
(3) Bars may be subdivided into parts of a whole or into percentages</td></tr>
<tr><td>Circle or Pie 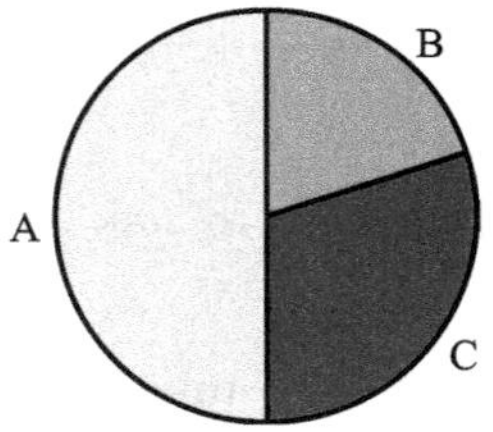</td><td>(1) It shows how various parts relate to a whole.
(2) It illustrates percentages</td></tr>
<tr><td>Solid Figure 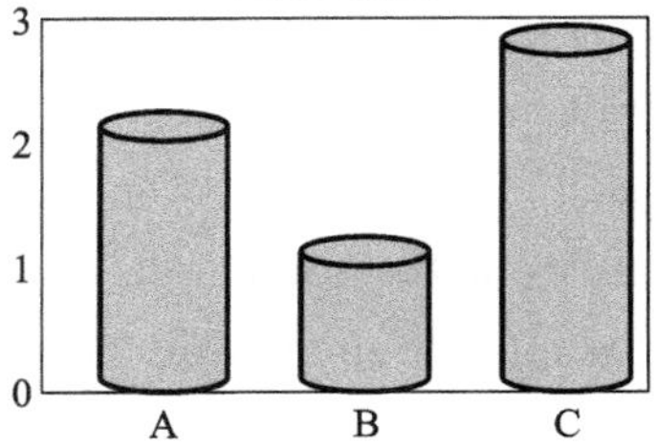</td><td>(1) It compares two or more totals using geometric figures to represent these quantities.
(2) Figures may be cubes, spheres, cylinders, etc</td></tr>
<tr><td>Picture or pictograph (Pictogram) 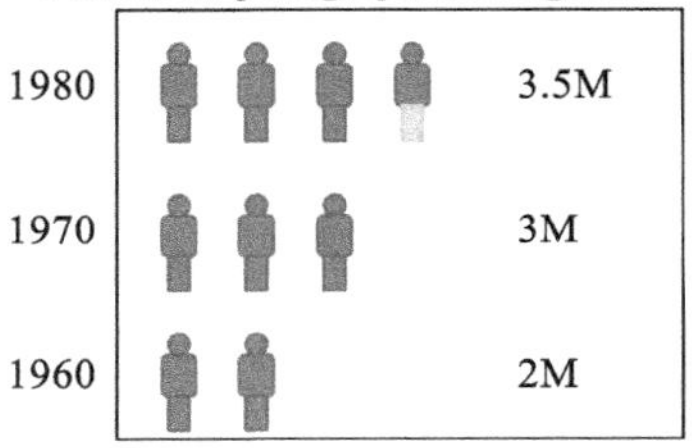</td><td>It illustrates approximate comparisons as bar graphs do, but uses representational figures, such as people, buses, cows, or other items being compared</td></tr>
</table>

TOPIC 8: Mechanical Manufacturing

Machinability of Metal Cutting

The term machinability, which is used in innumerable books, papers and discussions, may be taken to imply that there is a property or quality of a material which can be clearly defined and measured as an indication of the ease or difficulty with which it can be machined. In fact there is no clear cut unambiguous meaning to this term. To the active practitioner in machining, engaged in a particular set of operations, the meaning of the term is clear, and for him, machinability of a work material can often be measured in terms of the numbers of components produced per hour, the cost of machining the component, or the quality of the finish on a critical surface.

Problems arise because there are so many practitioners carrying out such a variety of operations, with different criteria of machinability. A material may have good machinability by one criterion, but poor machinability by another, or when a different type of operation is being carried out, or when conditions of cutting or the tool material are changed.

To deal with this complex situation, the approach adopted in this article is to discuss the behaviour of a number of the main classes of metals and alloys during machining, and to offer explanations of this behaviour in terms of their composition, structure, heat treatment and properties. The machinability of a material may be assessed by one or more of the following criteria.

(1) *Tool life*. The amount of material removed by a tool, under standardised cutting conditions, before the tool performance becomes unacceptable or the tool is worn by a standard amount.

(2) *Limiting rate of metal removal*. The maximum rate at which the material can be machined for a standard short tool life.

(3) *Cutting forces*. The forces acting on the tool (measured by dynamometer, under specified conditions) or the power consumption.

(4) *Surface finish*. The surface finish achieved under specified cutting conditions.

(5) *Chip shape*. The chip shape as it influences the clearance of the chips from around the tool, under standardised cutting conditions.

Magnesium

Of the metals in common engineering use, magnesium is the easiest to machine—the "best buy" for machinability, scoring top marks by almost all the criteria.

The tool forces when cutting magnesium are very low compared with those when cutting other pure metals, and they remain almost constant over a very wide range of cutting speeds, as shown in Figure 3-20. Both the cutting force (F_c) and the feed force (F_f) are low and the power consumption is considerably lower than that when cutting other metals under the same conditions. The low tool forces are associated with the low shear yield strength of magnesium, and, more importantly with the small area of contact with the rake face of the tool over a wide range of cutting speeds and rake angles. This ensures that the shear plane angle is high and the chips thin—only slightly thicker than the feed.

Aluminium

Alloys of aluminium in general also rate highly in the machinability table by most of the criteria.

In general, tool forces when cutting aluminium alloys are low, and tend to decrease slightly as the cutting speed is raised, as shown in Figure 3-20. High forces occur, however, when cutting commercially pure aluminium particularly at low speeds. In this respect aluminium behaves differently from magnesium, but in a similar way to many other pure metals. The area of contact on the rake face of the tool is very large, and this leads to a high feed force (F_f), low shear plane angle, and very thick chips, with consequent high cutting force (F_t) and high power consumption. The effect on pure aluminium of most alloying additions or of cold working, is to reduce the tool forces, particularly at low cutting speeds. In general most aluminium alloys, both cast and wrought, are easier to machine than pure aluminium, in spite of its low shear strength①.

EXERCISE 8: How to Read Graphs

Directions: Please answer the following questions based on Figure 3-20. The teacher and students can also ask their classmates or group members these questions to practice using reasoning skills to find information beyond the literal level.

(1) What kind of graph is Figure 3-20?

---

① shear strength 抗剪强度

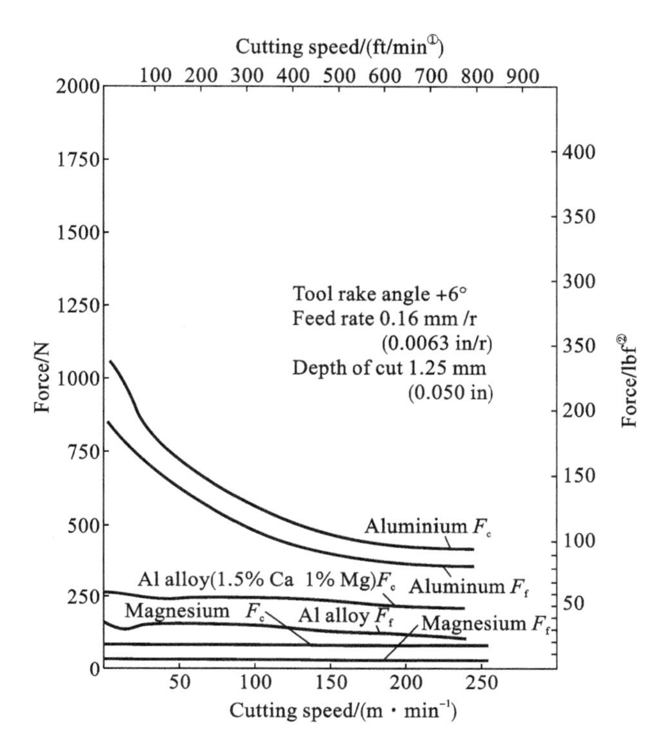

Figure 3-20 Tool Forces vs Cutting Speed—Magnesium and Aluminium (data From Williams, Smart and Milner 1)

(2) What is the main idea and purpose of the Figure 3-20?

(3) Write down the difficult words and try to find their meanings in a dictionary.

① 1 ft/min=0.304 m/min

② 1 lbf=4.45 N

(4) Identify the details that support the main idea.

(5) How can you reduce tool forces to cut aluminium?

(6) Which cutting speed range is recommended to cut aluminium and magnesium?

(7) New questions by teacher.

(8) New questions by students.

3.3.2 Types of Charts and How to Read Them

Charts are the summaries of important processes or relationships. They may combine pictorial, symbolic or verbal elements. Different types of charts are shown in Table 3-5.

Table 3-5 Different Types of Charts

Types of Charts	Description and Purpose	How to Read Charts
Flow 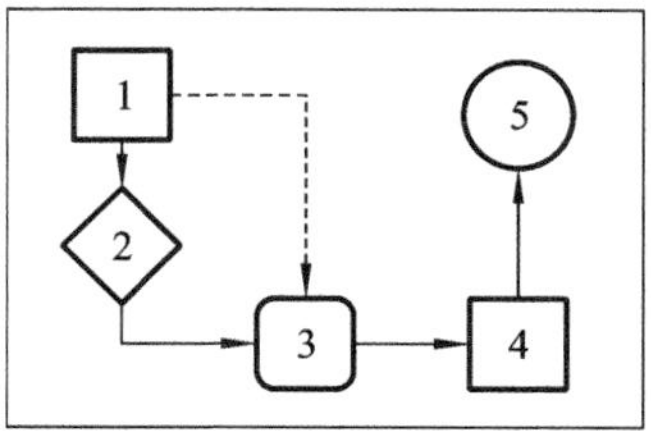 1 2 3 4 5	(1) It illustrates a process, functional relationship, organization. (2) It shows simple or complex sequences	(1) Note title and type of chart. They indicate main idea and purpose. (2) Note symbols. They should be easily recognized. Do not attempt to read them literally. Notice details. Observe relationships. (3) Note pattern of organization: cause and effect, comparison/contrast, chronology, classification, step-by-step procedure, system. (4) Make inferences and draw conclusions based on data. (5) Relate to text material
Tree 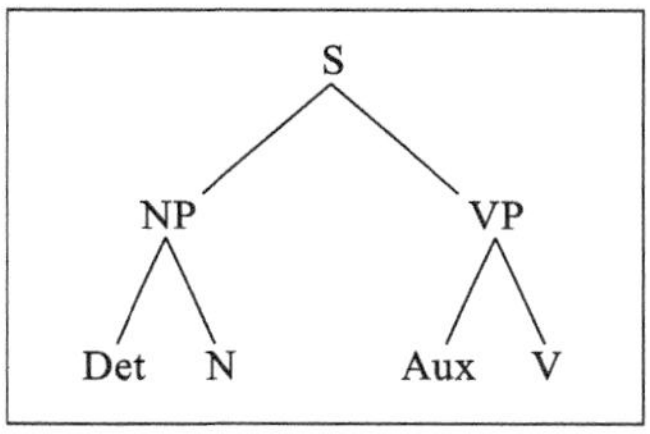 S NP VP Det N Aux V	(1) It shows the way in which many things developed from one source; depicts genealogies. (2) It shows development from root to many branches	
Time Line 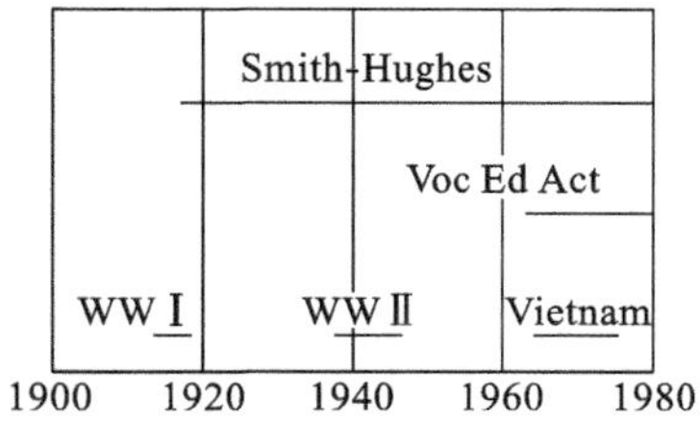 Smith-Hughes Voc Ed Act WW I WW II Vietnam 1900 1920 1940 1960 1980	(1) It shows relations among events. (2) It illustrates cause and effect, sequence. (3) Multiple lines may be used to show overlapping events	
Comparison Pro. Con. 1 1 2 2 3 3	(1) It compares and contrasts. Points may be listed side by side as advantages and disadvantages, pros and cons. (2) It may be verbal or statistical	
Diagram 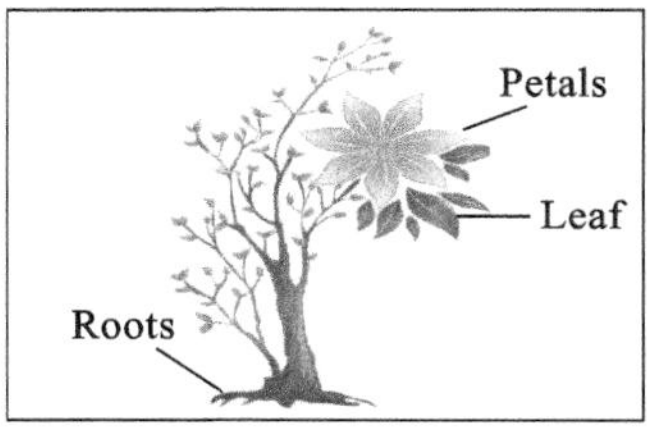 Petals Leaf Roots	(1) It shows structure of a system (schematic), steps in a process, parts of a structure. (2) It classifies complex procedures. There are many varieties—simple to complex	

TOPIC 9: Advanced Manufacturing

Artificial Intelligence

Artificial intelligence (AI) is that part of computer science concerned with systems that exhibit some characteristics usually associated with intelligence in human behavior (such as learning, reasoning, problem-solving, and the understanding of language). The goal of AI is to *simulate* such human behaviors on the computer. The art of bringing relevant principles and tools of AI to bear on difficult application problems is known as **knowledge engineering.**

Artificial intelligence is having a major effect on the design, the automation, and the overall economics of manufacturing operations, in large part because of advances in computer memory expansion (VLSI chip design) and decreasing costs. Artificial intelligence packages costing on the order of a few thousand dollars have been developed, many of which can be run on personal computers. Thus, AI has become accessible to office desks and shop floors.

Elements of Artificial Intelligence. In general, artificial intelligence applications in manufacturing encompass the following activities:

(1) expert systems;

(2) natural language;

(3) machine (computer) vision;

(4) artificial neural networks;

(5) fuzzy① logic.

1. Expert systems (ES)

An expert system (also called a knowledge-based system) is, generally, defined as an intelligent computer program that has capability to solve difficult real-life problems by the use of knowledge base and inference procedures. The goal of an expert system is the capability to conduct an intellectually demanding task in the way that a human expert would. The basic structure of an expert system is shown in Figure 3-21. The knowledge base consists of knowledge rules (general information about the problem) and the inference rules (the way conclusions are reached). The results may be communicated to the user through the natural-language interface.

The field of knowledge required to perform this task is called the domain of the

---

① fuzzy [ˈfʌzi] *adj.* 模糊的，不清楚的

expert system. Expert systems utilize a knowledge base containing facts, data, definitions, and assumptions. They also have the capacity for a heuristic[①] approach, that is, making good judgments on the basis of discovery and revelation, and making high-probability guesses, just as a human expert would.

The knowledge base is expressed in computer codes (usually in the form of if-then rules) and can generate a series of questions. The mechanism for using these rules to solve problems is called an inference engine. Expert systems can also communicate with other computer software packages.

To construct expert systems for solving the complex design and manufacturing problems encountered, one needs a great deal of knowledge and a mechanism for manipulating this knowledge to create solutions. Because of the difficulties involved in accurately modeling the many years of experience of an expert (or a team of experts), and the complex inductive reasoning and decision-making capabilities of humans (including the capacity to learn from mistakes), developing knowledge-based systems requires considerable time and effort.

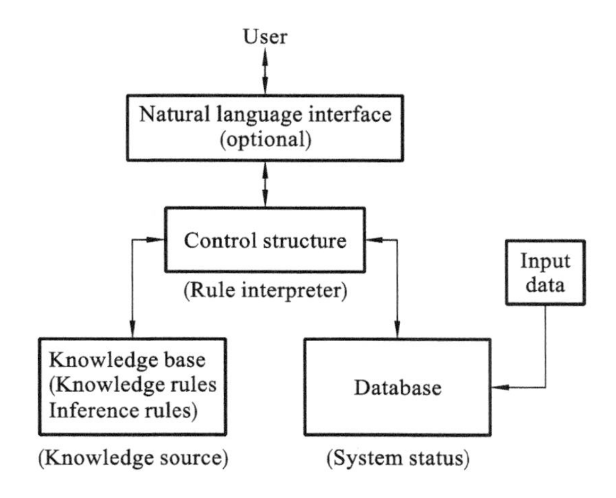

Figure 3-21 Basic Structure of an Expert System

Source: K. W. Goff, Mechanical Engineering, October 1985.

Expert systems operate on a real-time basis, and their short reaction times provide rapid responses to problems. The programming languages most commonly used for this application are C++, LISP and PROLOG; other languages can also be used. An important development is expert system software shells or environments (also called framework systems). These software packages are

① heuristic [hju'rıstık] *adj.* 启发式的

essentially expert-system outlines that allow a person to write specific applications to suit special needs. Writing these programs requires considerable experience and time.

Several expert systems have been developed and used since the early 1970s, ones can utilize computers with various capacities, such specialized applications are as the follows:

(1) problem diagnosis① in various types of machines and equipment, and determination of corrective actions;

(2) modeling and simulation of production facilities;

(3) computer-aided design, process planning, and production scheduling;

(4) management of a company's manufacturing strategy.

2. Natural Language Processing

Traditionally, obtaining information from a database in the computer memory required the utilization of computer programmers to translate questions in natural language into "queries②" in some machine language. Natural-language interfaces with database systems are in various stages of development. These systems allow a user to obtain information by entering English-language commands in the form of simple, typed questions.

Software shells are available, and they are used in such applications as the scheduling of material flow in manufacturing and the analyzing of information in databases. Significant progress is being made on computer software. They will have speech synthesis and recognition (voice recognition) capabilities to eliminate the need to type commands on keyboards.

3. Machine Vision

Computers and software implementing artificial intelligence are combined with cameras and other optical sensors. These machines then perform such operations as inspecting, identifying, sorting of parts, and guiding of robots (*intelligent robots*, as shown in Figure 3-22), operations that would otherwise require human intervention.

4. Artificial Neural Networks (ANN)

Although computers are much faster than the human brain at sequential③ tasks, humans are much better at pattern-based tasks that can be attacked with

---

① diagnosis [ˌdaɪəgˈnəʊsɪs] *n.* 诊断

② query [ˈkwɪəri] *n.* 问题，疑问

③ sequential [sɪˈkwenʃəl] *adj.* 按次序的，相继的，构成连续镜头的

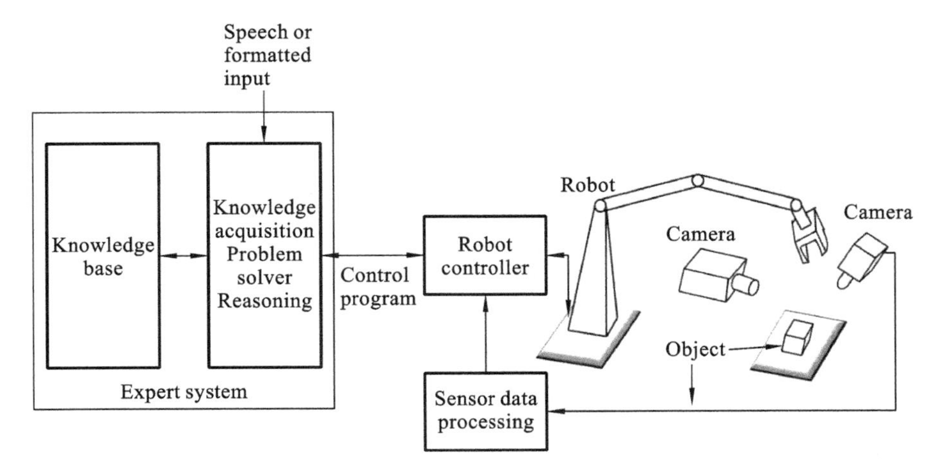

Figure 3-22 Expert System, as applied to an industrial robot guided by machine vision

parallel processing, such as recognizing features (in faces and voices, even under noisy conditions), assessing situations quickly, and adjusting to new and dynamic conditions. These advantages are also partly due to the ability of humans to use several senses (sight, hearing, smell, taste, and touch) simultaneously (data fusion①) and in real time. The branch of artificial intelligence called artificial neural networks attempts to gain some of these capabilities through computer imitation of the way that data is processed by the human brain.

The human brain has about 100 billion linked neurons② (cells that are the fundamental functional units of nervous tissue③) and more than a thousand times that many connections. Each neuron performs only one, simple task: it receives input signals from a fixed set of neurons and, when those input signals are related in a certain way (specific to that particular neuron), it generates an electrochemical④ output signal which goes to a fixed set of neurons. It is now believed that human learning is accomplished by changes in the strengths of these signal connections between neurons.

A fully developed, feed-forward network is the most common type of ANN, and it is built according to this principle from several layers of processing elements (simulating neurons). The elements in the first (input) layer are fed with input

---

① fusion [ˈfjuːʒən] *n.* 联合，合并

② neuron [ˈnjʊərɒn] *n.* 神经元，神经细胞

③ tissue [ˈtɪsjuː] *n.* 组织

④ electrochemical [ɪˌlektrəʊˈkemɪkəl] *adj.* 电化学的

data— for example, forces, velocities, voltages[①]. Each element sums up all its inputs: one per element in the input layer, many per element in succeeding layers. Each element in a layer then transfers the data (according to a transfer function) to all the elements in the next layer. Each element in that next layer, however, receives a different signal, because of the different connection weights between the elements.

The last layer is the output layer, within which each element is compared to the desired output—that of the process being simulated. The difference between the desired output and the calculated one (the error) is fed back to the network by changing the weights of the connections in a way that reduces this error. After this procedure has been repeated several times, the network has been "trained", and it can now be used on input data not previously presented to the system.

Other kinds of ANN are associative memories, self-organizing ANN, and adaptive-resonance[②] ANN. The feature common to these neural networks is that they must be trained with concrete[③] exemplars[④]. It is, therefore, very difficult to formulate input-output relations mathematically and to predict an ANN's behavior with untrained inputs.

Artificial neural networks are being used in such applications as noise reduction (in telephones), speech recognition, and process control. For example, they can be used for predicting the surface finish of a workpiece obtained by end milling, on the basis of input parameters such as cutting force, torque[⑤], acoustic[⑥] emission, and spindle[⑦] acceleration. Although still controversial, the opinion of many is that true artificial intelligence will evolve only through advances in ANN.

5. Fuzzy Logic

An element of artificial intelligence having important applications in control systems and pattern recognition is fuzzy logic (fuzzy models). Introduced in 1965, and based on the observation that people can make good decisions on the basis of imprecise and non-numerical information, fuzzy models are mathematical means of representing vagueness[⑧] and imprecise information (hence, the term "fuzzy").

---

① voltage ['vəʊltɪdʒ] *n.* 电压，伏特数

② resonance ['rezənəns] *n.* 共振，调谐

③ concrete ['kɒŋkriːt] *adj.* 实际的，具体的，明确的

④ exemplar [ɪg'zemplɑː] *n.* 模型，模范

⑤ torque [tɔːk] *n.* （力）转矩，力矩，扭矩

⑥ acoustic [ə'kuːstɪk] *adj.* 声音的，听觉的

⑦ spindle ['spɪndl] *n.* 轴

⑧ vagueness ['veɪgnəs] *n.* 暧昧，含糊

These models have the ability to recognize, represent, manipulate, interpret, and utilize data and information that are vague or lack precision. These methods deal with reasoning and decision-making at a level higher than do neural networks. Typical linguistic examples are the following: few, very, more or less, small, medium, extremely, and almost all.

Fuzzy technologies and devices have been developed (and successfully applied) in areas such as robotics and motion control, image processing and machine vision, machine learning, and the design of intelligent systems. Some applications are in the automatic transmission of the Lexus automobile; a washing machine that automatically adjusts the washing cycle for load size, fabric type, arid amount of dirt; and a helicopter that obeys vocal commands to go forward, up, left, and right, to hover, and to land.

EXERCISE 9: How to Read Charts

Directions: Please answer the following questions based on Figure 3-21 & Figure 3-22. The teacher and students can also ask their classmates or group members these questions to practice using reasoning skills to find information beyond the literal level.

(1) What information can be understood in Figure 3-21?

________ Illustrates cause and effect.

________ Shows functional relationship organization.

________ Shows simple or complex sequences.

________ Shows structure of a system.

(2) What information can be understood in Figure 3-22?

________ Illustrates cause and effect.

________ Shows functional relationship organization.

________ Shows simple or complex sequences.

________ Shows structure of a system.

(3) The organization pattern of Figure 3-21 & Figure 3-22 are ______________ and ________________.

(4) What information does Figure 3-22 supply for text material?

(5) New questions by teacher.

(6) New questions by students.

3.3.3 Types of Illustrations and How to Read Them

Different types of illustrations are shown in Table 3-6.

Table 3-6 Types of Illustrations

<table>
<tr><th>Types</th><th>Description</th><th>Purpose</th><th>How to Read Illustrations</th></tr>
<tr><td>Photograph</td><td>(1) The most realistic two-dimensional illustration.
(2) May be abstract, however.
(3) May be distorted due to selection, point of view and/or editing</td><td>(1) Generates interest. Motivates and clarifies text by providing a sense of reality.
(2) Effectively shows step-by-step procedures, comparisons, status of things, processes, scenes, events, people</td><td rowspan="3">(1) Read the caption. This may be misleading. If so, is there a reason?
(2) Survey illustration, get general impression.
(3) Look for details—objects, colors, symbols. Notice relationships.
(4) Make inferences, draw conclusions, seek applications.
(5) Refer to text in order to determine how to read illustration in terms of author's purpose for using it.
(6) Repeat step 4</td></tr>
<tr><td>Painting, Drawing</td><td>(1) Wide variety in types and styles.
(2) The most realistic portrayal[①] of pre-photography eras.
(3) Often found in history, literature, and psychology texts.
(4) May provide interpretation of a scene, event, person.
(5) May represent abstract ideas, feelings, emotions</td><td>(1) Effectively demonstrates values, styles, and concerns of an era, group of persons or individuals.
(2) Compares and contrasts values, styles, and concerns of eras, groups of persons, individuals.
(3) It is a kind of visual poetry</td></tr>
<tr><td>Cartoon</td><td>(1) Compact pictorial representation of ideas, employing caricature, symbolism, exaggerative humor, satire.
(2) Variety of artistic techniques.
(3) Frequently biased, distorted views.
(4) May employ symbols, which can become outdated because they usually are related to events, styles, thoughts of the era in which they appear.
(5) Relies on stereotypes.
(6) Purpose must be perceived in order to read the meaning</td><td>(1) Gains attention.
(2) Illustrates ideas, opinions. Criticizes, satirizes, prophesies.
(3) Introduce humor.
(4) Induced self-examination, self-criticism</td></tr>
</table>

① portrayal [pɔːˈtreɪəl] *n.* 描绘；画像，肖像

TOPIC 10: Robots

Advanced Vision Algorithm[①] Helps Robots Learn to See in 3D

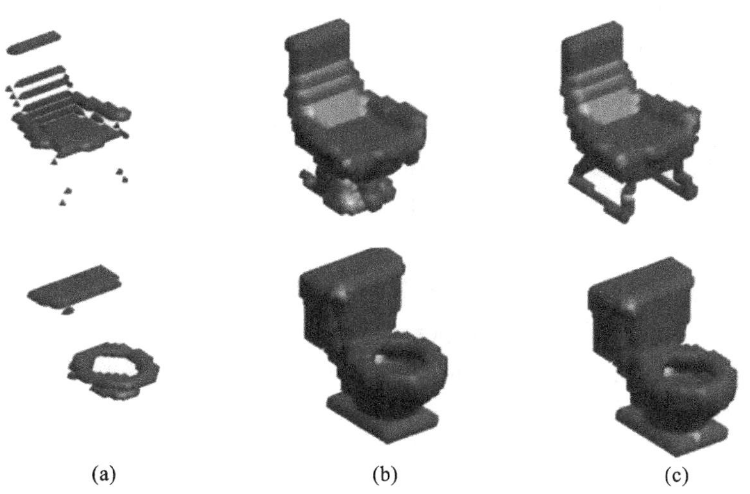

Figure 3-23 The Ability of a Robot to Recognize Three-dimensional Objects

Credit: Ben Burchfield

Robots are reliable in industrial settings, where recognizable objects appear at predictable time in familiar circumstances. But life at home is messy. Put a robot in a house, where it must navigate unfamiliar territory cluttered with foreign objects, and it's useless.

Now researchers have developed a new computer vision algorithm that gives a robot the ability to recognize three-dimensional (3D) objects and, at a glance, intuit items that are partially obscured or tipped over, without needing to view them from multiple angles. As shown in Figure 3-23, when fed 3D models of household items in bird's eye view Figure 3-23 (a), a new algorithm is able to guess what the objects are, and what their overall 3D shapes should be. This image shows the guess model (Figure 3-23 (b)), and the actual 3D model (Figure 3-23 (c)).

"It sees the front half of a pot sitting on a counter and guesses there's a handle in the rear and that might be a good place to pick it up from," said Ben Burchfiel, a Ph. D. candidate in the field of computer vision and robotics at Duke University.

In experiments where the robot viewed 908 items from a single vantage point, it guessed the object correctly about 75 percent of the time. State-of-the-art

① algorithm [ˈælgərɪðəm] *n.* [计][数] 算法，运算法则

computer vision algorithms previously achieved an accuracy of about 50 percent.

Burchfiel and George Konidaris, an assistant professor of computer science at Brown University, presented their research at the Robotics: Science and Systems Conference in Cambridge, Massachusetts.

RELATED: Personalized Exoskeletons are Making Strides toward a Man-Machine Interface

Like other computer vision algorithms used to train robots, their robot learned about its world by first sifting through a database of 4000 3D objects spread across ten different classes — bathtubs, beds, chairs, desks, dressers, monitors, night stands, sofas, tables, and toilets.

While more conventional algorithms may, for example, train a robot to recognize the entirety of a chair or pot or sofa or may train it to recognize parts of a whole and piece them together, this one looked for how objects were similar and how they differed.

When it found consistencies within classes, it ignored them in order to shrink the computational problem down to a more manageable size and focus on the parts that were different.

For example, all pots are hollow in the middle. When the algorithm was being trained to recognize pots, it didn't spend time analyzing the hollow parts. Once it knew the object was a pot, it focused instead on the depth of the pot or the location of the handle.

"That frees up resources and makes learning easier," said Burchfiel.

Extra computing resources are used to figure out whether an item is right-side up and also infer its 3D shape, if part of it is hidden. This last problem is particularly vexing in the field of computer vision, because in the real world, objects overlap.

To address it, scientists have mainly turned to the most advanced form of artificial intelligence, which uses artificial neural networks, or so-called deep-learning algorithms, because they process information in a way that's similar to how the brain learns.

Although deep-learning approaches are good at parsing complex input data, such as analyzing all of the pixels in an image, and predicting a simple output, such as "this is a cat," they're not good at the inverse task, said Burchfiel. When an object is partially obscured, a limited view — the input — is less complex than the output, which is a full, 3D representation.

The algorithm Burchfiel and Konidaris developed constructs a whole object from partial information by finding complex shapes that tend to be associated with each other. For instance, objects with flat square tops tend to have legs. If the robot can only see the square top, it may infer the legs.

"Another example would be handles," said Burchfiel. "Handles connected to cylindrical drinking vessels tend to connect in two places. If a mug shaped object is seen with a small nub visible, it is likely that nub extends into a curved, or square, handle."

RELATED: Construction Robot can "Print" a Building in 14 Hours

Once trained, the robot was then shown 908 new objects from a single viewpoint. It achieved correct answers about 75 percent of the time. Not only was the approach more accurate than previous methods, it was also very fast. After a robot was trained, it took about a second to make its guess. It didn't need to look at the object from different angles and it was able to infer parts that couldn't be seen.

This type of learning gives the robot a visual perception that's similar to the way humans see. It interprets objects with a more generalized sense of the world, instead of trying to map knowledge of identical objects onto what it's seeing.

Burchfiel said he wants to build on this research by training the algorithm on millions of objects and perhaps tens of thousands of types of objects.

"We want to build this into single robust system that could be the baseline behind a general robot perception scheme," he said.

Originally published on Seeker.

EXERCISE 10: How to Read Illustrations

Directions: Please answer the following questions based on TOPIC 10. The teacher and students can also ask their classmates or group members these questions to practice using reasoning skills to find information beyond the literal level.

(1) What is your first impression of TOPIC 10?

(2) Why the models in the center are slightly different to models on the right?

(3) What is the main purpose of using Figure 3-23?

(4) What is the new computer vision algorithm?

(5) New questions by teacher.

(6) New questions by students.

3.4 Reading Experience Ⅳ: Developing Integral Technical Reading Skills

3.4.1 Reading and Analyzing the Materials

Reading the Materials

The first step that of carefully reading the materials to be assigned seems obvious. However, instructors who have been using the same materials for several years sometimes fail to reread them each time they use them. They may forget what information is actually there.

But reading the materials is not enough. How you read the materials is perhaps the most important aspect in reading, because through careful analysis you will determine what is to learn and what skills you need to learn.

Analyzing the Materials

In analyzing instructional materials you are, essentially, asking a series of questions about the materials.

Concepts

(1) What are the important concepts?

(2) If the concept is a process, is there a chart or diagram that presents the process visually? If so, should the focus of the initial discussion be on the chart or diagram?

(3) What topics are covered? (List them as you read. Often headings and subheadings are provided.)

Vocabulary

(1) What technical terms are used to describe the concepts and processes?

(2) What are the relationships between and among these words? That is, do the meanings of some words have to be known first because they become essential components in the definitions of other terms?

(3) How are the words presented in the materials?

(4) What would be the best way to learn these words?

Organization

(1) How is the information presented? That is, how does the author organize the materials? Is he/she primarily comparing or contrasting? Is he/she providing instruction in how to do something using a step-by-step procedure?

(2) How much and what kind of background do you need to have in order to understand the information?

(3) Does the author provide clear signal words as clues to organizational patterns?

TOPIC 11: Mechatronics

What is and not mechatronics?

Mechatronics is special. It is no more a mere mixture of electronics, mechanics, and computing than a *Chateau Latour* (or *Grange Hermitage*) vintage wine is a mixture of yeast and grape juice.

Long ago, Karel Čapek wrote a book, Rossum's Universal Robots. It was as little about robotics as animal farm was about agriculture, but the term had been coined. Science fiction writers grew fat on the theme, and the idea of mechanical slave workers was lodged in the mind of the public.

When Devol designed a mechanical manipulator, Unimate, it was endowed with the term "a robot arm". As a research topic, robotics ceased to be about tin men and turned to the articulation of mechanical joints to move a gripper or workpiece to a precise set of coordinates. The new "Three Laws of Robotics" concerned the Denavit-Hartenberg transformation matrices, discrete-time control algorithms, and precision sensors.

Robotics is just a narrow subset of mechatronics. It is true that it has all the ingredients of sensing, actuation, and a quantity of computer-assisted strategy in between, but with every day the list of mechatronic products increases. In video recorders, DVD players, jet airliners, fuel injection motor engines, advanced sewing machines, and Mars rovers, not to mention all the gadgetry that surrounds a computer, the jigsaw① pieces of mechatronics are slotted together.

In something as simple as a thermostat②, sensing and actuation of the heater are linked. But the element of computation is missing. It is not mechatronic. In automatic sliding doors, however, the criterion is not as cut and dried. A few simple logic circuits are enough to link the passive infrared sensor③ to the door motor, but the designer might have found that the alternative of embedding a microprocessor was in fact simpler to design and cheaper to construct.

---

① jigsaw ['dʒɪgsɔː] *n.* 拼图玩具；线锯；镂花锯

② thermostat ['θɜːməstæt] *n.* 恒温器；自动调温器

③ passive infrared sensor 被动式红外传感器

Before 1960, autopilots were capable of automatic landing. Their computational processes were based on *magnetic amplifiers*①, circuits using the saturation of a mumetal core with no semiconductor② more complicated than a diode③. As the aircraft approached its target, the mode switching from height lock to ILS (instrument landing system) radio beam to flareout④ controlled by a radar altimeter⑤ was performed by a clunking Ledex switch, a rotary solenoid⑥ driving something similar to an old radio waveband changer.

This must come close to qualifying as robotics, but lacking any trace of digital computation, it must fall short of mechatronics. For today's aircraft, however, with digital autopilots that can not only guide the aircraft across the world and land it, but also taxi it to the selected air bridge at the terminal, there can be no question that it is a mobile robot.

Machines that can roll, walk, climb, and fly under their own automatic control have come to share the title of robots, mobile robots. One example of such a robot is the Micromouse. IEEE Spectrum Magazine and David Christiansen must take the credit for devising a contest in which small trolleys explore a maze. You may like to claim personal credit for redefining the maze design and rules to give victory to the "intelligent" mouse, rather than the "dumb wall followers".

Many early Mice used stepper motors to move and steer them, controlled by microprocessors of one sort or another. The maze walls were sensed by a variety of photoelectric devices, although in at least two cases mechanical "feelers⑦" were used with great success. To navigate through the maze, a map had to be built up in the microcomputer's memory. To solve the maze, a strategy was required. A further aspect of the software was the need to apply control to keep the mouse straight as it ran through the passageways. So, in one not-so-simple contest, all the ingredients of mechatronics were brought together.

The contest runs regularly to this day. Many of the early champions are still at the forefront, while simplified versions of the contest have been developed to encourage young entrants. While the experts hone their expertise, however, the

---

① magnetic amplifier 磁放大器

② semiconductor[ˌsemikənˈdʌktə] *n.* [电子][物] 半导体

③ diode [ˈdaɪəʊd] *n.* [电子] 二极管

④ flareout[ˈfleəaʊt] *n.* [航] 拉平；扩展；均匀；扩口

⑤ radar altimeter [雷达] 雷达高度计

⑥ rotary solenoid [电磁] 旋转螺线管

⑦ feeler [ˈfiːlə] *n.* 试探者；探测器，探针，触针，探头

bar has to be set lower and lower for the newcomers. Simply running through a twisted path with no junctions is a testing problem for most schools' entrants.

So, what is the "mechatronic approach"? How would a mechatronics engineer design a set of digital bathroom scales? Would they be based on a straingauge① sensor, on the "twang②" frequency of a wire tensioned by the user's weight, or on some more subtle piece of ingenuity③?

When you opened up the machine on your bathroom floor, you may be disappointed to discover that the pointer of a conventional mechanical scale had simply been replaced with a disk with a notched edge. As it rotated under the weight of the user, an incremental optical encoder counted the notches of the disk as they went by and displayed the count on a luminous④ display.

For a manufacturing company with an established market in mechanical scales, the "pasted on" digital feature makes sense. However a "truly mechatronic" solution would find a tradeoff between digits and mechanical precision that would simplify the product.

A hairdryer marketed some years ago featured a "bonnet", coupled by a hose to the hot-air unit. A plastic knob could be rotated to give continuously variable temperature control. So, how would you go about designing it? When the question is put to university classes, it always brings answers featuring potentiometers, thyristor⑤ power controllers, and often a microcomputer.

The product was actually much simpler. The airflow was divided into two paths after the fan. In one path there was a heating element, regulated by a simple thermostat just "downstream", while the other simply blew cold air. The ornate knob moved a shutter that closed off one or other flow, or allowed a variable mixture of the two.

Good design can often demand an awareness of how to avoid excessive technology.

EXERCISE 11: Reading and Analyzing the Materials

Directions: Please answer the following questions based on TOPIC 11. The teacher and students can also ask their classmates or group members these

---

① straingauge *n.* 应变仪

② twang [twæŋ] *n.* 鼻音；弦声；砰然一声

③ ingenuity [ˌɪndʒəˈnjuːəti] *n.* 心灵手巧，独创性；精巧；精巧的装置

④ luminous [ˈluːmɪnəs] *adj.* 发光的；明亮的；清楚的

⑤ thyristor [ˈθaɪrɪstə] *n.* [电子] 半导体闸流管

questions to practice using reasoning skills to find information beyond the literal level.

(1) What is mechatronics?

(2) How can one distinguish if a system is a mechatronical system or not?

(3) What technical terms are used to describe the concept of mechatronics?

(4) How does the author organize the material?

________ Compare or contrast

________ Provide instruction of step-by-step procedure

________ Time sequence

________ Tell about the cause and effect

________ Listing

________ Categorization

(5) What kind of background of Devol's design and Micromouse do you need in order to understand the information?

(6) New questions by teacher.

(7) New questions by students.

3.4.2 Using a Text Analysis Chart

You might find it helpful to use a simple chart as an aid in analyzing instructional materials. As you prepare to read, a text analysis chart forces you to organize your thinking by providing space for recording the various elements of the reading. There should be space in the chart for noting the following.

- **Major concepts or ideas**—Make sure the really important ideas in the reading that you want to be sure of are not missed. For example, an abstract principle (e. g. supply and demand), a formula, a law of physics, or an essential operating procedure.
- **Topics covered in the reading**—Describe the way in which the subject matter is organized, divided, and sequenced.
- **Essential vocabulary**—Record the new words you want to know at the end of the lesson or the unit.
- **Organizational pattern**—Note how the author goes about providing the information (e. g. comparison, cause and effect, or time sequence).
- **Reading/study skills needed**—List any particular skills one must have in order to gain from the reading assignment. The skills may include such things as reading graphics, using a book index, or analyzing words.
- **Important visual aids**—This entry may include the visuals in the materials themselves, visuals in other books or learning guides, or other visual aids you need to help understand the reading.

TOPIC 12: Quality Control and Inspection

The Need for Quality Control

The major goals for quality control are manifold and include a need to determine the real functional tolerances required for engineering specifications and to provide a plan for control of the quality of the results of a process related to time. This is normally achieved by statistical or control chart methods.

Additionally, it is essential to know when the set-up of the equipment is safe enough to permit a production run that should not produce defective parts beyond an acceptable limit. It is also important to obtain variations in the products arising from the inherent variability of the following factors:

(1) the fabricating equipment,

(2) the material,

(3) the operator.

Variations must be small enough to meet product specification requirements (the combined inherent product variation is called the *process capability*①).

It is also necessary to be able to confirm, through various sampling plans or inspection methods, that the process quality is controlled and that the inspection process is as economical, as a balance between the risks involved will permit.

Finally, a major goal must be the ability to improve the quality of performance of any processes or product designs, using a statistical analysis of its variability.

It is also worth remembering that, for a good commercial design, a specification should be based on "the lowest acceptable quality for the minimum practical cost". This quality must take into account customers' reaction and safety requirements.

If all the above goals are realized, then the instigation of quality control can be claimed a success. However, a feature of paramount importance is that of economy.

Improved quality may increase the value and cost of the product (Figure 3-24) and therefore initiating quality control in an organization must be carried out efficiently. The areas where control is required must be highlighted and investigated. They may be identified in the following ways:

(1) customers' complaints;

(2) large amounts of rejects;

(3) large amounts of rework;

---

① process capability 加工能力

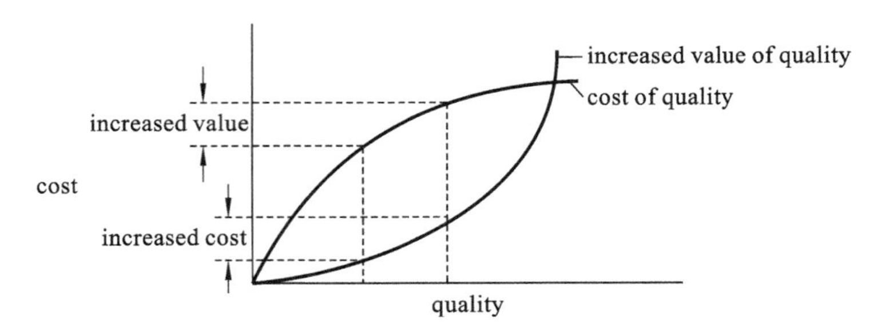

Figure 3-24 Increased Cost and Value of Product as a Function of Quality

(4) late arrivals of components and products, or late deliveries;

(5) deviation requests which may be raised, these should be made difficult to raise;

(6) analysis of inspection labour force to determine its effectiveness.

Control considerations

When considering control, the following questions should be asked.

(1) What is the value of likely scrap[①] against the cost of control?

(2) What level of inspection is appropriate (sample or 100%)?

(3) What type of inspection scheme is appropriate, discrete or variable?

(4) How frequent should inspection be? This is decided by the likelihood of failure (per unit time), how serious are the consequences of failure (value added at later stages of manufacture, or danger to life), and the cost of inspection (Figure 3-25).

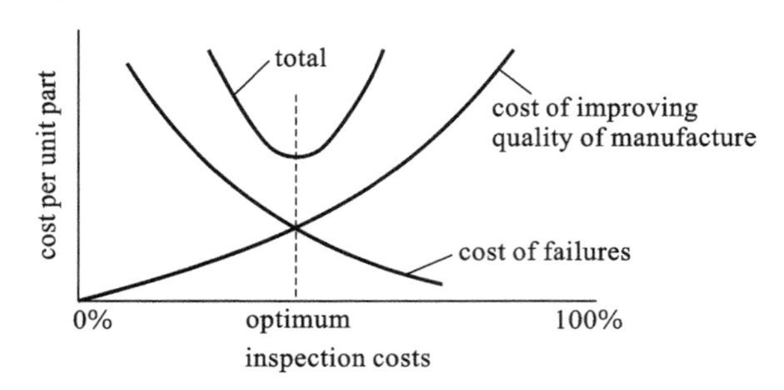

Figure 3-25 Frequency of Inspection and its Consequences, in Terms of Cost of Inspection

(5) Where are the critical areas of inspection? For instance, in suppliers' organizations, at manufacture, or during assembly, during storage or after transit.

(6) Is it feasible to link operator pay to quality, thus making them responsible for it?

① scrap [skræp] *n.* 碎片；残余物；少量

(7) What is the realistic deployment of inspectors, and what is the required level of training for them?

(8) Should improved efficiency of inspection performance be related to pay?

EXERCISE 12: Using a Text Analysis Chart

Directions: This part of the exercise is consistent with the method in 3. 4. 1. Fill the chart based on TOPIC 12.

Text Analysis Chart

<table>
<tr><td colspan="4">Reading Materials: ____________________</td></tr>
<tr><td colspan="2">Major Concepts
1.
2.
3.
4.
…</td><td colspan="2">Content Topics (as listed in text)</td></tr>
<tr><td colspan="2">Essential Vocabulary</td><td colspan="2">Organizational Pattern</td></tr>
<tr><td colspan="2">Reading/Study Skills Needed</td><td colspan="2">Important Visual Aids</td></tr>
<tr><td colspan="4">Resources for Background and Motivation</td></tr>
<tr><td>People</td><td>Places</td><td>Media</td><td>Others</td></tr>
<tr><td></td><td></td><td></td><td></td></tr>
</table>

Chapter 4　Learning Professional English Anytime, Anywhere

4.1　Ad from Magazine

The Figure 4-1 is an ad from Internet—A Perfectly-Balanced Hard Hat For Welders. Please try to consider these questions.

(1) What is the main purpose of welding helmet①?

(2) What is the existing problem of helmet during welding?

(3) What is the greatest advantage of this hard hat for welders?

You could also use the methods mentioned in Chapter 3 to find out what information does this ad attempt to send to people?

4.2　News—The World's Lightest Material

News are written clearly and directly, making them perfect for language learning. The news can be your own personal English classroom, especially when the news contains technical knowledge.

Please read the following news about the world's lightest material and fill in the form in EXERCISE 12.

It's incredibly light, yet amazingly strong and rigid, as shown in Figure 4-2.

It's called microlattice②, the lightest metal structure ever created, and it was created by scientists at HRL Laboratories, the same Malibu, California-based company that invented the laser and the semiconductor.

HRL Labs is jointly owned by Boeing and General Motors and scientists believe microlattice could revolutionize the way airplanes, cars and even buildings are constructed.

This month, Popular Mechanics magazine named the microlattice as one of 10 World-Changing Innovations for 2012.

---

① welding helmet *n.*　焊工帽；焊工面罩

② microlattice *n.*　微格金属；微晶格；迷你网格；微格

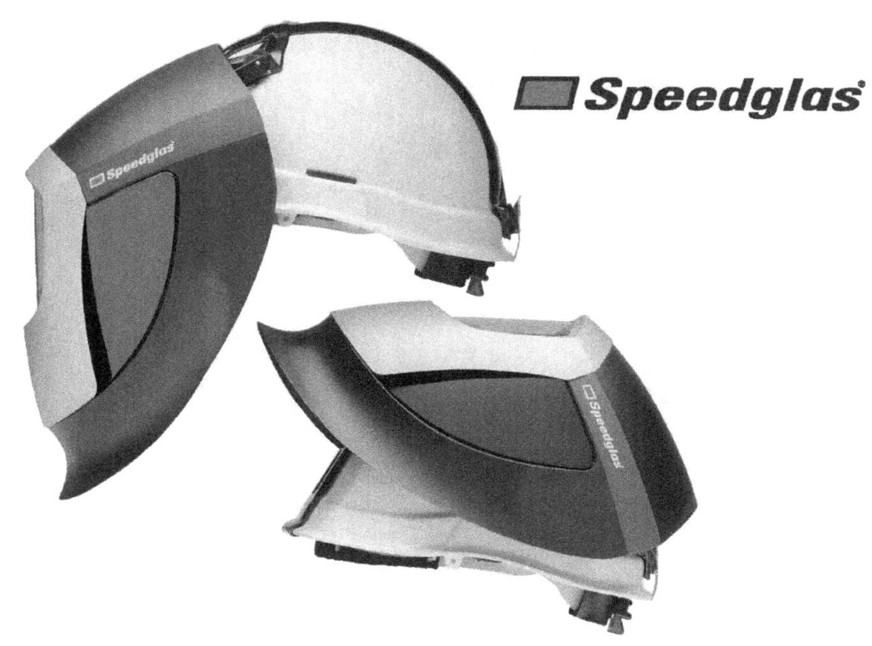

A Perfectly-Balanced Hard Hat For Welders

The new Speedglas® ProTop has an exclusive pivot mechanism that allows the welding helmet to "nest" down low onto the hard hat, in a perfectly-balanced, locked-down position.

Its unique design eliminates the neck strain caused by traditional welding hard hats, which have very high centers-of-gravity when raised. The nested helmet is also ideal for moving around obstacles and tight spaces. The helmet automatically unlocks when the welder pulls it back down into position over the face.

Plus:

- Four auto-darkening Speedglas® welding lenses to choose from.
- Aerodynamically-designed exhaust vents for helmet "breathability."
- Optional Speedglas SideWindows™ increase peripheral vision by 100%.
- Independently tested to meet ANSI Z89.1 (head protection) and ANSI Z87.1 (eye & face protection).

HÖRNELL

HORNELL, INC.
2374 EDISON BLVD. • TWINSBURG, OH 44087 USA
TEL: 800-628-9218 • 330-425-8880 • FAX: 330-425-4576
EMAIL: info.us@hornell.com • www.hornell.com

©2002 Hornell. Speedglas®, Fresh-air® and SideWindows™ are trademarks of Hornell. Protected by patents worldwide.

Figure 4-1 An Ad from Internet

Figure 4-2　The Lightest Material

"We're creating the next-generation of lightweight materials and what we're trying to do is to incorporate design into the material itself," said Alan Jacobsen, microlattice inventor and senior scientist at HRL Labs. "So just like you'd designed a skyscraper or the frame of the 787, here we're designing the material. You decide what properties you want, and then we design the materials for the particular properties."

For Boeing, the creation of the ultra light metallic microlattice, from a polymer template, opens the door for numerous potential uses, including lightweight airplane construction, acoustic① management and thermal absorption.

"This invention may allow us to construct a one-piece, single structure unmanned aerial vehicle②," said Gail Taylor-Smith, Boeing HRL Technical Director. "With mechanical properties that allow the substance to flex, we might one day be able to compress space systems to reduce deployed volume, but once in orbit could be expanded to normal size, like a pop-up satellite."

Additionally, because the microlattice can be manipulated and produced and reproduced quickly, it's optimal for rapid manufacturing.

"We could also have mass customization③ with this material and fabricate directly on the line," Taylor-Smith said, "It will be easy and low-cost to make adjustments on the fly. It's amazing."

---

① acoustic [ə'kuːstɪk] *adj.* 声学的；音响的；听觉的

② unmanned aerial vehicle 无人驾驶飞行器

③ mass customization 大规模定制

4.3 How to Read a Function/Formula

4.3.1 How to Read Simple Symbols and Equations

Common pronunciations (in British English-Gimson, 1981) of mathematical and scientific symbols are given in the Table 4-1 and Table 4-2.

1. Symbols

Table 4-1 Symbols and Pronunciations

Symbol	Pronunciation	Symbol	Pronunciation
$+$	plus	$f^{(4)}$	f four; fourth derivative
$-$	minus	∂	partial derivative, delta
$\pm$	plus or minus	$\int$	integral
$\times$	multiplied by	$\sum$	sum
$/$	over; divided by	w. r. t.	with respect to
$\div$	divided	log	log
$=$	equals	$\log_2 x$	log to the base 2 of x
$\approx$	approximately, similar	$\therefore$	therefore
$\equiv$	equivalent to; identical	$\because$	because
$\neq$	not equal to	$\rightarrow$	gives, leads to, approaches
$>$	greater than	$/$	per
$<$	less than	$\in$	belongs to; a member of; an element of
$\geqslant$	greater than or equal to	$\notin$	does not belong to; is not a member of; is not an element of
$\leqslant$	less than or equal to	$\subset$	strictly contained in; a proper subset of
$\not>$	not greater than	$\subseteq$	contained in; subset
$\not<$	not less than	$\cap$	intersection
$\gg$	much greater than	$\cup$	union
$\ll$	much less than	$\forall$	for all
$\perp$	perpendicular to	$\cos x$	cos x; cosine x

Continued Table 4-1

Symbol	Pronunciation	Symbol	Pronunciation
$\parallel$	parallel to	$\sin x$	sine x
$\not\equiv$	not equivalent to, not identical to	$\tan x$	tangent x
$\not\simeq \not\approx$	not similar to	$\csc x$	cosecant x
^{2}	squared	$\sinh x$	shine x
^{3}	cubed	$\cosh x$	cosh x
^{4}	to the fourth; to the power four	$\tanh x$	than x
^{n}	to the nth; to the power n	$\lvert x \rvert$	mod x; modulus x
$\sqrt{\ }$	root; square root	℃	degrees Centigrade
$\sqrt[3]{\ }$	cube root	℉	degrees Fahrenheit
$\sqrt[4]{\ }$	fourth root	°K	degrees Kelvin
!	factorial	0 °K, −273.15 ℃	absolute zero
%	percent	mm	millimetre
∞	infinity	cm	centimetre
$\propto$	varies as; proportional to	cm^3	cubic centimetre, centimetre cubed
·	dot	m	metre
··	double dot	km	kilometre
:	is to, ratio of	mg	milligram
$f(x)$ fx	f; function	g	gram
$f'(x)$	f dash; derivative	kg	kilogram
$f''x$	f double-dash; second derivative	AC	A. C.
$f'''(x)$	f triple-dash; f treble-dash; third derivative	DC	D. C.

2. Examples of expressions and equations/formulas

Table 4-2 Examples and Pronunciations

Example	Pronunciation
$x+1$	x plus one
$x-1$	x minus one
$x\pm1$	x plus or minus one
xy	x y; x times y; x multiplied by y
$(x-y)(x+y)$	x minus y, x plus y
x/y	x over y; x divided by y;
$x\div y$	x divided by y
$x=5$	x equals 5; x is equal to 5
$x\approx y$	x is approximately equal to y
$x\equiv y$	x is equivalent to y; x is identical with y
$x\neq y$	x is not equal to y
$x>y$	x is greater than y
$x<y$	x is less than y
$x\geqslant y$	x is greater than or equal to y
$x\leqslant y$	x is less than or equal to y
$0<x<1$	zero is less than x is less than 1; x is greater than zero and less than 1
$0\leqslant x\leqslant 1$	zero is less than or equal to x is less than or equal to 1; x is greater than or equal to zero and less than or equal to 1
x^2	x squared
x^3	x cubed
x^4	x to the fourth; x to the power four
x^n	x to the nth; x to the power n
x^{-n}	x to the minus n; x to the power of minus n
$\sqrt{x}$	root x; square root x; the square root of x
$\sqrt[3]{x}$	the cube root of x
$\sqrt[4]{x}$	the fourth root of x
$\sqrt[n]{x}$	the nth root of x
$(x+y)^2$	x plus y all squared
$(x/y)^2$	x over y all squared
$n!$	n factorial; factorial n

Continued Table 4-2

Example	Pronunciation
$x\%$	x percent
∞	infinity
$x \propto y$	x varies as y; x is (directly) proportional to y
$x \propto \frac{1}{y}$	x varies as one over y; x is indirectly proportional to y
$\dot{x}$	x dot
$\ddot{x}$	x double dot
$f(x)$ fx	f of x; the function of x
w. r. t.	with respect to
$\log_e y$	log to the base e of y; log y to the base e; natural log (of) y
$\therefore$	therefore
$\because$	because
m/s	metres per second
$x \in A$	x belongs to A; x is a member of A; x is an element of A
$x \notin A$	x does not belong to A; x is not a member of A; x is not an element of A
$A \subset B$	A is strictly contained in B; A is a proper subset of B
$A \subseteq B$	A is contained in B; A is a subset of B
$A \cap B$	A intersection B
$A \cup B$	A union B
$\cos x$	cos x; cosine x
$\sin x$	sine x
$\tan x$	tangent x, tan x
$\csc x$	cosecant x
$\sinh x$	shine x
$\cosh x$	cosh x
$\tanh x$	than x
$\|x\|$	mod x; modulus x
18 ℃	eighteen degrees Centigrade
70 ℉	seventy degrees Fahrenheit

4.3.2 How to Read Symbols and Equations in Calculus

Calculus is a unique branch of mathematics, and it includes many symbols and equations that are also unique. Some are intuitive and make sense at a glance, but others can be very confusing when you are not instructed on what they mean. Here is a quick overview of some of the symbols you will come across in calculus.

1. $\lim\limits_{x \to x_0} f(x)$

This is the format for writing a limit in calculus. When read aloud, it says "The limit of the function f of x, as x tends to x_0."

2. $f'(x)$

This is a common symbol indicating the derivative of the function $f(x)$. It reads simply as "The derivative of f of x."

3. dy/dx

This is another symbol for a derivative. You can read it as "The derivative of y with respect to x." y is equivalent to $f(x)$, as y is a function of x itself.

4. $f''(x)$, d^2y/dx^2

Both of these symbols represent the second derivative of the function, which means you take the derivative of the first derivative of the function. You would read it simply as "The second derivative of f of x."

5. $f^n(x)$, d^ny/dx^n

These symbols represent the nth derivative of $f(x)$. Much like the second derivative, you would perform differentiation on the formula for n successive times. It reads as "The nth derivative of f of x." If n were 4, it would be "The fourth derivative of f of x." for example.

6. $\int_a^b f(x)$

This symbol represents the integration of the function. The integration of the function is essentially the opposite of the differentiation. The variables a and b represent the lower limit and upper limit of the section of the graph the integral is being applied to. If there are no values for a and b, it represents the entire function. You would read it as "The integral of f of x with respect to x (over the domain of a to b)."

7. $\dot{y}$

This is the symbol for differentiation with respect to time. You can read it as "the derivative of y with respect to time."

This is just a small sample of the symbols of equations involved in calculus, but should provide a decent launching point for being able to understand calculus symbols and equations.

Other symbols of equations in calculus are shown in Table 4-3.

Table 4-3 Symbols of Equations in Calculus and Pronunciation

Symbol	Pronunciation
∂v	the partial derivative of v
$\frac{\partial v}{\partial \theta}$	delta v by delta θ, the partial derivative of v with respect to θ
$\frac{\partial^2 v}{\partial \theta^2}$	delta two v by delta θ squared; the second partial derivative of v with respect to θ
$\mathrm{d}v$	the derivative of v
$\int$	integral
$\int_0^\infty$	integral from zero to infinity
$\sum$	sum
$\sum_{i=1}^{n}$	the sum from i equals 1 to n
$\rightarrow$	gives, approaches
$\Delta x \rightarrow 0$	delta x approaches zero
$\lim_{\Delta x \rightarrow 0}$	the limit as delta x approaches zero, the limit as delta x tends to zero

4.4 How to Read Engineering Drawing

When learning this part, it seems quite daunting[①] because there is so much information to take in. This unit will break down a drawing/print and make it easier to understand. But, first thing is first! What is the purpose of an engineering drawing if there are 3D models? Engineering drawings show numerous features of a part that a 3D model doesn't. For instance, drawings show the material type, the finish[②], dimensions[③], hardware[④], company information, and other specific requirements. The sole purpose of a drawing is to show all the details of a part. Imagine if you were looking at a single part in your hand, a drawing would essentially describe and illustrate all the details of how to place the part in your hand.

Engineering drawings can be referred to manufacturing prints, dimensional prints, prints, drawings, manufacturing blueprints[⑤], blueprints, mechanical drawings, and more. Drawings are designed by engineers, so there are a lot of engineering lingoes[⑥] and symbols that are used to identify and describe certain aspects of a part. This is the "daunting" part of a drawing because unless you have experience in reading drawings or you're an engineer, learning what everything means on a drawing can be a challenge. So let's break down a drawing and make it easier! This unit will specify the format, location, and type of information that should be included in drawings like **blocks, notes, specifications, and symbols** you may find on a drawing.

4.4.1 Information Blocks

Engineers include a lot of critical information in blocks in order to give the reader information about who the drawing belongs to, part number and description, and information about the material and finish. The information blocks are located in the bottom right-hand corner of the drawing. Occasionally, some blocks are left

① daunting [ˈdɔːntɪŋ] *adj.* 使人畏缩的；使人气馁的；令人怯步的
② finish [ˈfɪnɪʃ] *n.* 表面处理
③ dimension [dɪˈmenʃən] *n.* 规模，大小
④ hardware [ˈhɑːdweə] *n.* 计算机硬件；五金器具
⑤ blueprint [ˈbluːprɪnt] *n.* [印刷] 蓝图；模板
⑥ engineering lingo 工程术语

blank if the information in that block is not needed or hasn't been decided. From a manufacturing point of view, the more information the better and all blocks should be filled in so there are no assumptions. The information blocks include the followings.

1. Title Block

The following information is located within the title block in the lower right hand corner of a drawing as shown in Figure 4-3.

- Name: company or agency who prepared or owns the drawing
- Address: location of the company or agency
- Name and date: responsible engineers who drew, checked, and approved the drawing
- Part name/description: describes what the part is
- Part/drawing number: assigned number to identify the part
- Revision: identifies the correct version of the drawing
- Scale (optional): ratio of actual size of the part compared to the size of the part on the drawing. It can be shown as 1 : 1 or 1=1. The first number represents the actual size of the part and the second number represents the drawing. In other words, 1 : 2 means the drawing is double the actual size. Whereas 3 : 1 indicates the actual size is three times what is shown on the drawing. Note: even if the scale says 1 : 1, never measure the drawing, use the dimensions
- Size: specifies the drawing sheet size; A = 8.5×11, B = 11×17, C = 17×22, D = 22×34, E = 34×44, F = 28×40 (the unit is the inch)

2. Revision Block

In addition to the revision shown in the title block, there is a revision block (Figure 4-4) that is located in the upper right hand corner. The revision block indicates the specifics in regards to the changes that were made to roll the revision. The revision block includes the revision, the description of what changes were made, the date of the revision, and the approval of the revision.

3. Bill of Materials (BOM) Block

For parts that require assembly, require hardware to be added to the part, or there is a kit of parts, there is a bill of materials block that contains a list of all the items that are needed for the assembly or project. Bill of materials can also be referred to as a parts list, schedule, or for short BOM (pronounced either "be-oh-em" or "bomb"). The BOM includes the part number, the description/name of the part or item, the material specification (if any), and the quantity to be required of that item.

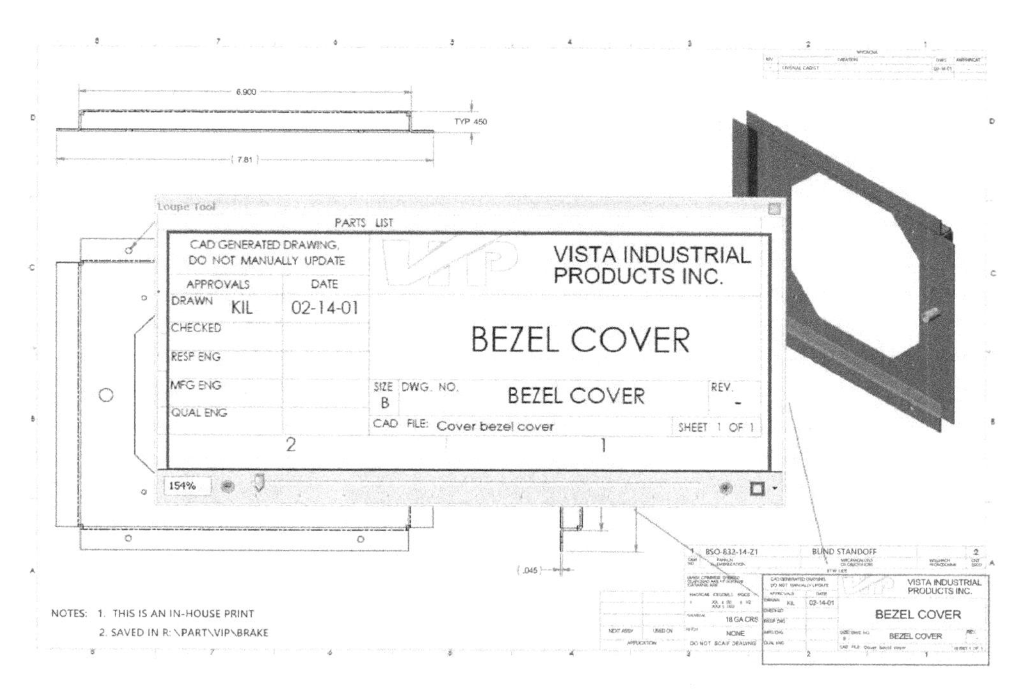

Figure 4-3 Title Block

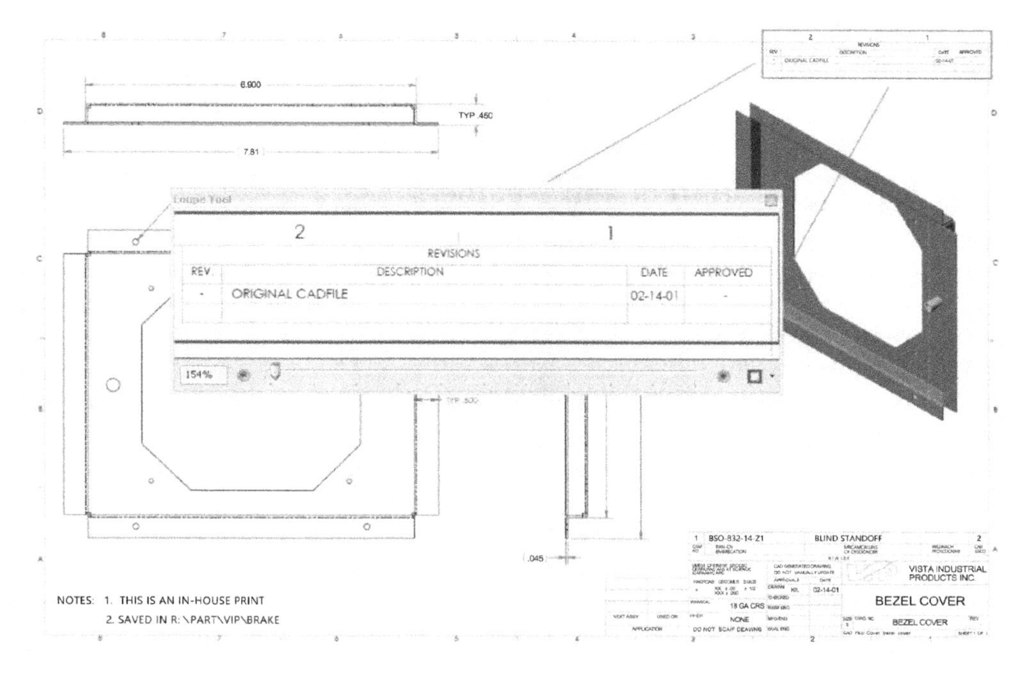

Figure 4-4 Revision Block

The BOM can be located in different areas on the print. The two most common areas are located right above the title block or in the upper left hand corner (Figure 4-5).

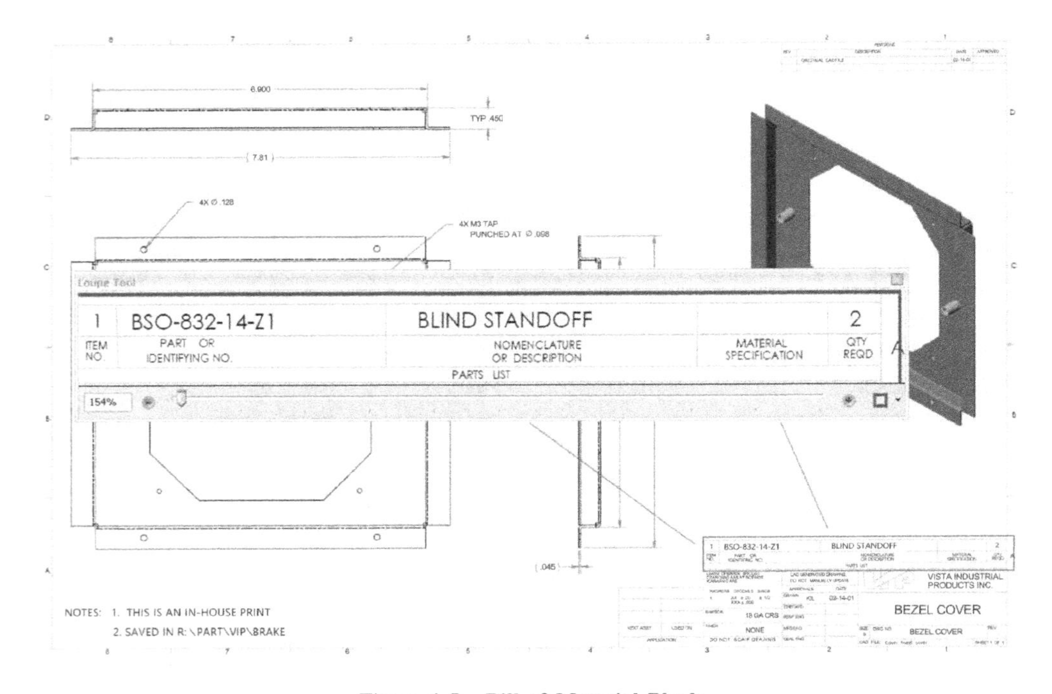

Figure 4-5 Bill of Material Block

4.4.2 Zone Letters and Numbers

Print contains a border around it which includes letters and numbers as shown between the margin in Figure 4-6. These letters and numbers are used like a map in order to help locate and pinpoint certain areas of a print. The letters are in an alphabetical order from the bottom up. The numbers are in a numerical order from right to left. Therefore, you are to read the zones from right to left.

4.4.3 Notes and Specifications

Since the print shows the part graphically and with dimensions, sometimes there are many specifics to the part that cannot be seen graphically which are described in the notes and specifications which are generally located in the upper left hand corner or lower left hand corner (Figure 4-7).

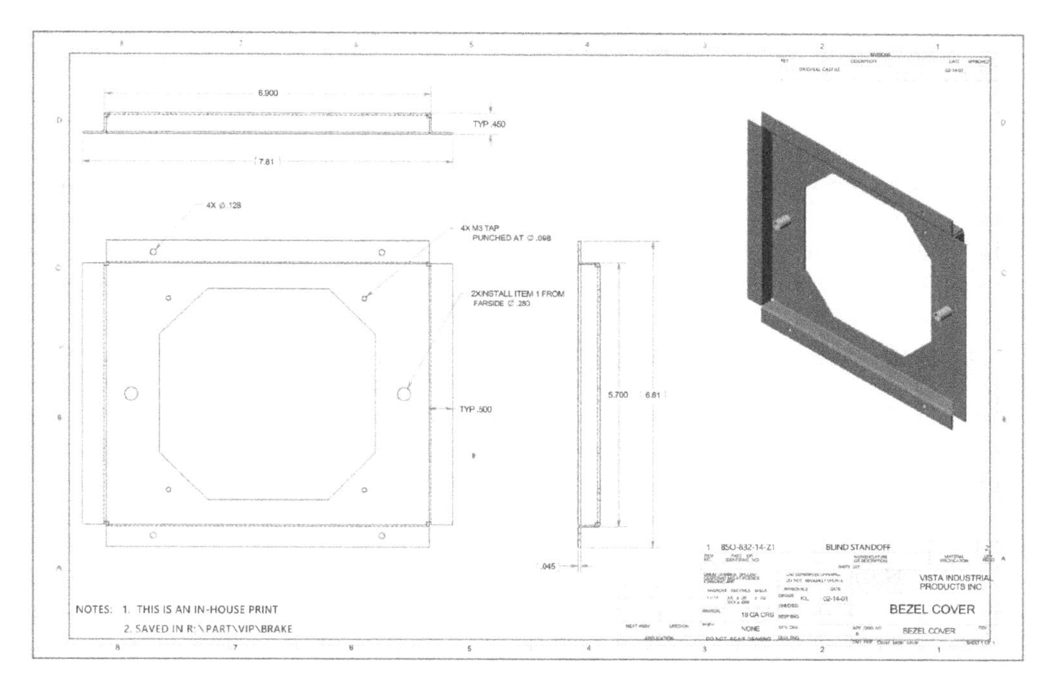

Figure 4-6 A Print with Border

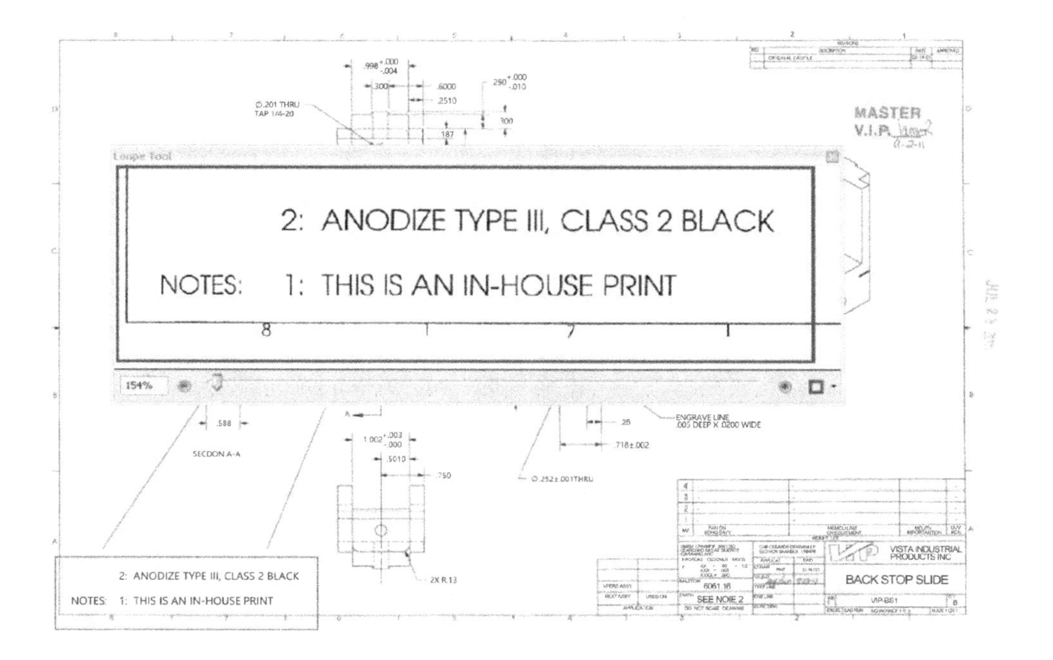

Figure 4-7 Note of a Print

Notes are additional information about the part. Whereas, **specifications** are a reference to an actual document or statement that describes how the parts are to be manufactured, assembled, and maintained.

Now that all the different components of a print have been identified and described, the next step is learning how to read a manufacturing print. The following part will help you to understand the lines of a part of a print in order to visualize① the part correctly.

4.4.4 Lines on a Drawing

Once there is an understanding of the different components of a print and where everything is located, the next step is to be able to read the lines on a print. Reading a print means to understand what the graphic of a part is showing. Therefore, you must understand how lines work on a print. For engineers and manufacturers, lines are their communicators or even their alphabet which convey information. Table 4-4 is a chart of the various lines that are used on a print and their descriptions.

By understanding these lines and how they are used, it will be much easier for you to look at a print and be able to visualize how the part would actually look.

Table 4-4 Lines and Descriptions

Name	Graphic	Description
Object Line	———————	Solid lines used to form the shape of a part
Hidden Line	- - - - - - - - - - -	Dashed lines used to form the shape of features that are not visible
Center Line	——— · ——— · ———	Solid line with a dash then a solid line used to identify the center of a feature
Dimension Line	←—— ——→	Solid line with arrowhead tips followed by a dimension, used to identify where the dimension represents
Leader Line	————————→	Solid line with arrowhead tip that is associated with a note or specification pointing to a location on the part, but not with a dimension

① visualize [ˈvɪʒuəlaɪz] *vt.* 使形象化；想象，设想

Continued Table 4-4

Name	Graphic	Description
Break (long)	—⋀⋁—⋀⋁—⋀⋁—	Thin solid line with zigzag used to reduce the size of part to show entire object and reduce details
Break (short)	~~~~~~	Thick solid wavy line used to indicate a short break
Section Line	\\\\\\\\\\\\\\\\\\\\\\\\\\\\\\	Multiple diagonal lines used to indicate the surface in the section view to have been cut along the cutting-plane line
Cutting-plane Line	┌—— ——┐	U-shaped line with arrowhead tips, used to designate an imaginary cut
Phantom Line	—··—··—··—··—	Series of one long dash, two short dashes, and a long dash used to show an alternate position of a part

4.4.5 Some Examples of Engineering Drawing

Engineering drawing is primarily used to provide information about an object rather than a pleasing image. It must convey enough information to create or use the object in the physical world. Figure 4-8 to Figure 4-15 are some simple examples about engineering drawing, which only provide part of information needed to describe an object.

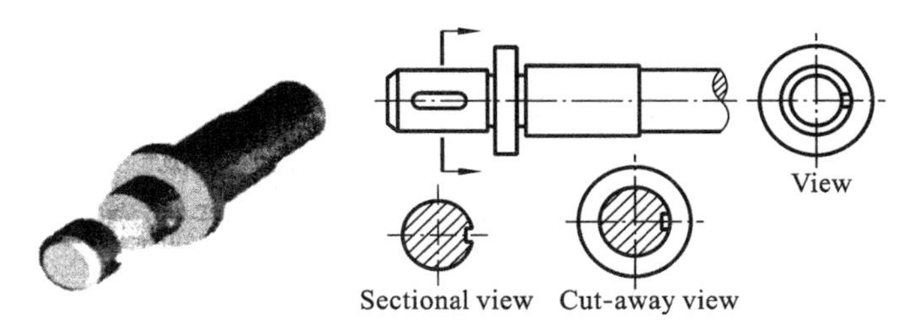

Figure 4-8 Comparison Between Sectional View and Cut-away View

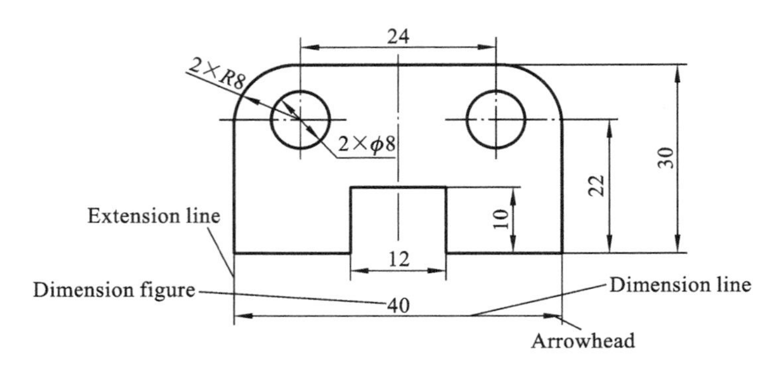

Figure 4-9 Components of Dimension

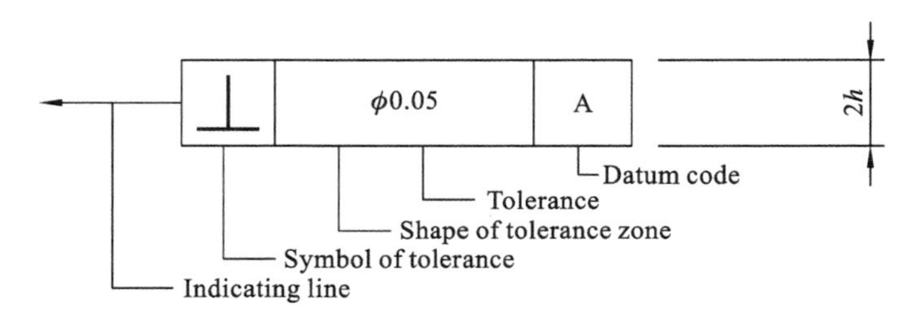

Figure 4-10 Code for Shape and Location Tolerance

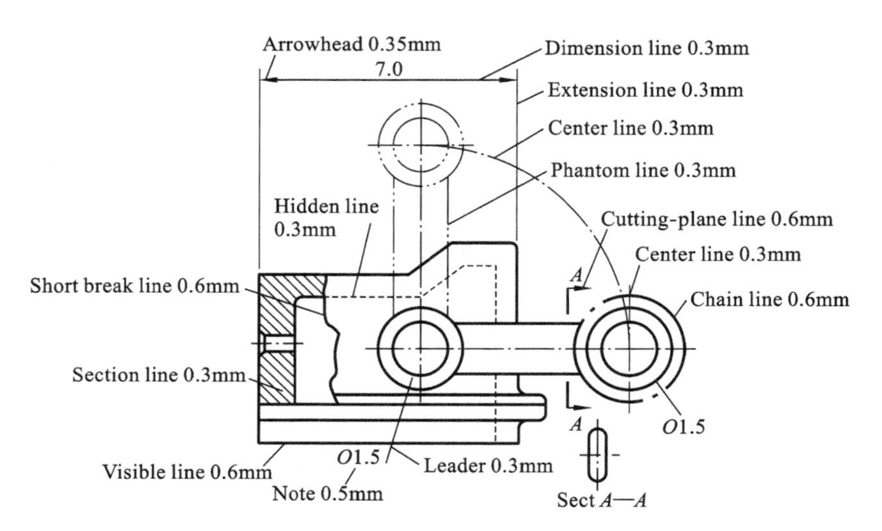

Figure 4-11 Application of Different Line Types

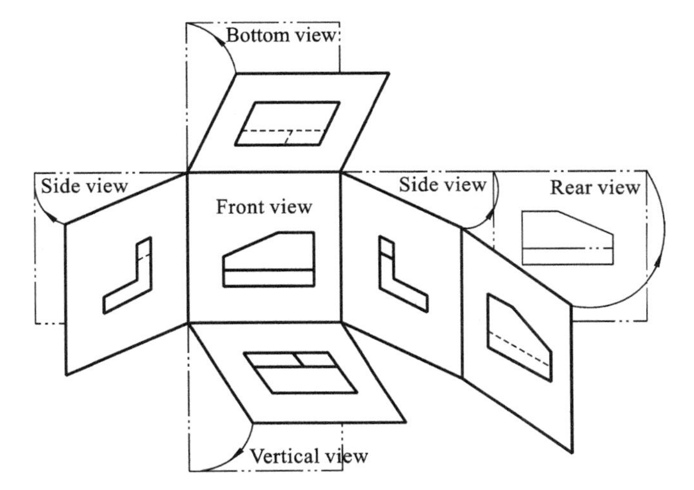

Figure 4-12 Developed Representation of the Six Views

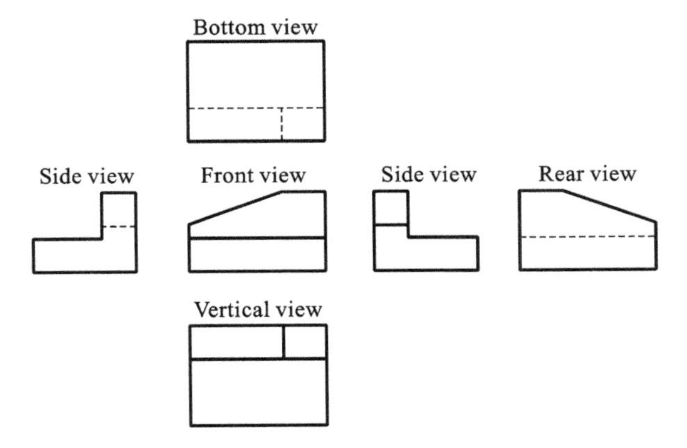

Figure 4-13 Location of Six Views

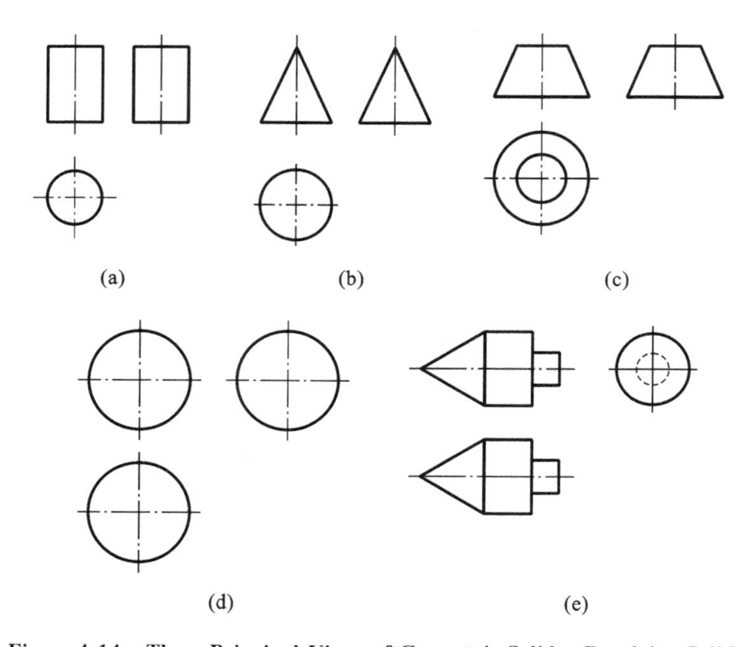

Figure 4-14 Three Principal Views of Geometric Solids: Revolving Solids

(a) Cylinder (b) Cone (c) Frustum (d) Sphere (e) Coaxial revolving solid

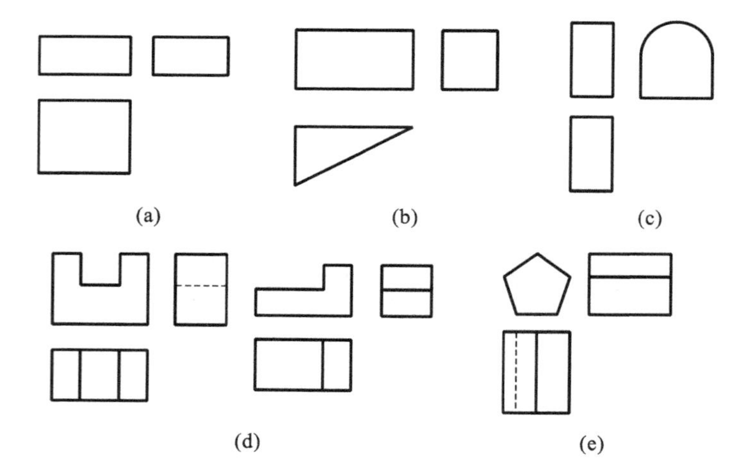

Figure 4-15 Three Principal Views of Geometric Solids: Polyhedra

(a) Cuboid (b) Wedge (c) U-shaped cylinder (d) Solid of extrusion (e) Right pentagonal prism

Chapter 5 English in Experiments

This chapter will start with an interesting experiment to show you how to understand the simple experiment instructions from daily life.

5.1 Interesting Experiment—Mechanical Hand Build up Mission

These project-based lessons focus on basic principles of physics, structural, and mechanical engineering. Physical models are built from a similar set of materials that can be easily sourced online.

Mission

(1) You will make a simple mechanical hand with a trigger① connected to a hinge②(Figure 5-1).

(2) Once complete, you can test the ability of your hand by trying to pick up as many straws as possible. Bring it to the class and make a competition about straws picking!

(3) Then make a presentation about the mechanism of motion.

Steps

1. Materials

For this design, you will need(and extra for redesigning): 15 craft sticks, 8 craft cubes, 4 round cube beads (larger pack size), part of 1 straw, 2 dowels of 0.4 mm, 2 rubber bands③, tape, hot glue④, and a box of straws for the challenge!

2. Build the Hinge

Cut two 1/2-inch⑤ pieces of straw. Wrap tape around the end of a skewer and thread the round cube beads and the straw pieces onto it as shown in Figure 5-2.

---

① trigger [ˈtrɪgə] *n.* 扳机；[电子] 触发器；制滑机

② hinge [hɪndʒ] *n.* 铰链，折叶；关键，转折点；枢要，中枢

③ rubber band 橡皮筋

④ hot glue 热溶胶；热黏接剂

⑤ 1 inch=2.54 cm

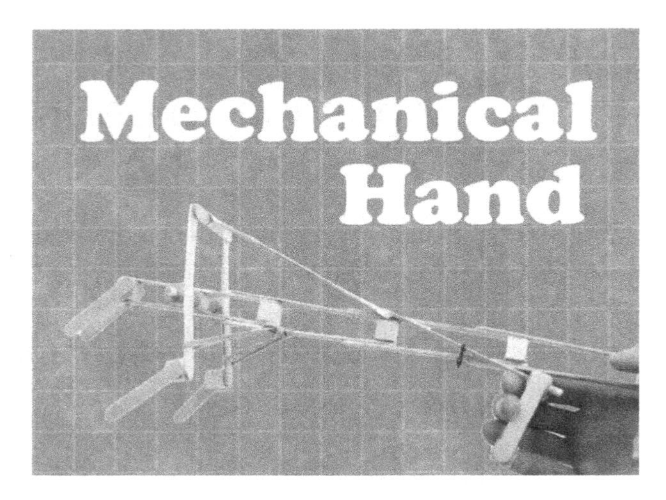

Figure 5-1 Mechanical Hand

Wrap tape around the other side and cut off the excess.

The straw pieces act as spacers① to prevent the fingers from colliding②.

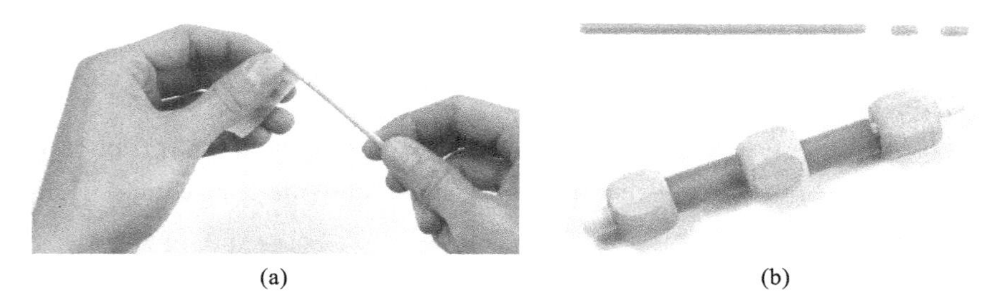

(a) (b)

Figure 5-2 Build the Hinge

3. Make the Fingers

Create the center finger as shown in Figure 5-3. The half-sticks are glued onto the side of the cube at a 45° angle.

The two other fingers are built similarly on either side of the center finger. Make sure that the half-sticks on the outside fingers are glued on at an angle that is inverse to the center finger.

On the "back" of the hand, glue a half-stick onto the two outside fingers. These two fingers will be actuated at the same time.

4. Make the Arm

The arm is an extension of the center finger. Sticks are glued together with at

① spacer [speɪsə] *n.* 间隔器

② colliding [kəˈlaɪdɪŋ] *v.* 碰撞；冲突

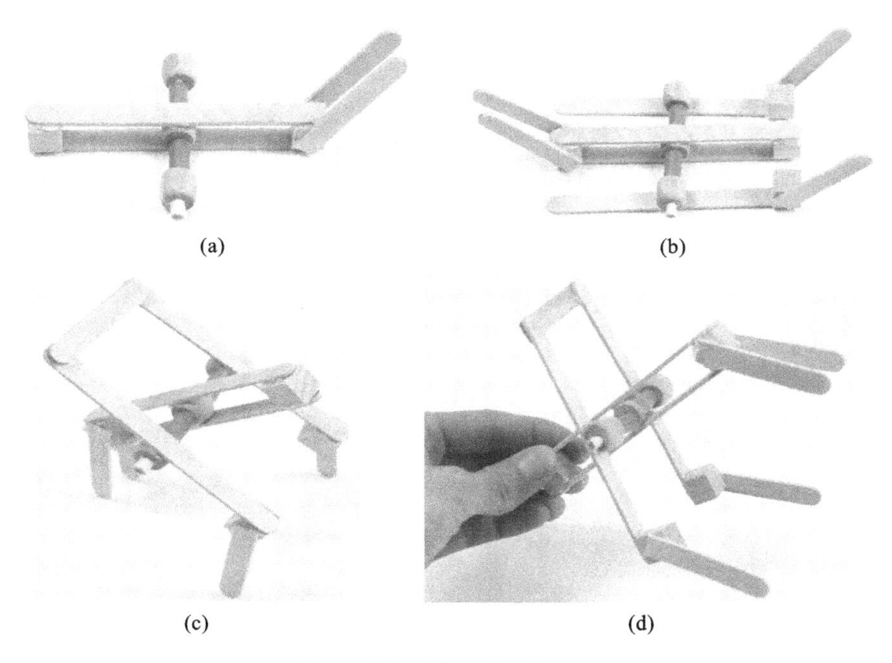

(a) (b) (c) (d)

Figure 5-3 Make the Fingers

least 1-inch of overlap to ensure a nice strong bond. The arm can be any length, but longer than 6 or 7 sticks is not recommended because it is more difficult to lift objects that are further away from your body. The arm also may not be able to support its own weight if it is extremely long.

Once the arm length is determined, add a handle and a thumb rest at the end. The handle and thumb rests that are pictured in Figure 5-4 are very minimal—there is a lot of room for innovation and customization.

5. The Trigger

A trigger is something that activates a mechanism—it's not a term just used for operating a gun. The trigger is what transfers the movement from the user's hand to the mechanical hand.

Place two skewers end to end such that the pointed ends are facing each other. Firmly wrap a 10 cm piece of tape lengthwise around the skewers. Attach one end to the "back" of the hand with several layers of tape as shown in the Figure 5-5.

Make the trigger as shown in Figure 5-5(b). Thread it onto the skewer.

The mechanical hand needs to be calibrated to match the user's finger length. Hook your thumb around the thumb rest and place the handle against the base of your thumb. Extend the other 4 fingers and place the trigger just under the first digit of your hand. Wrap tape around either side of the trigger to hold it in place

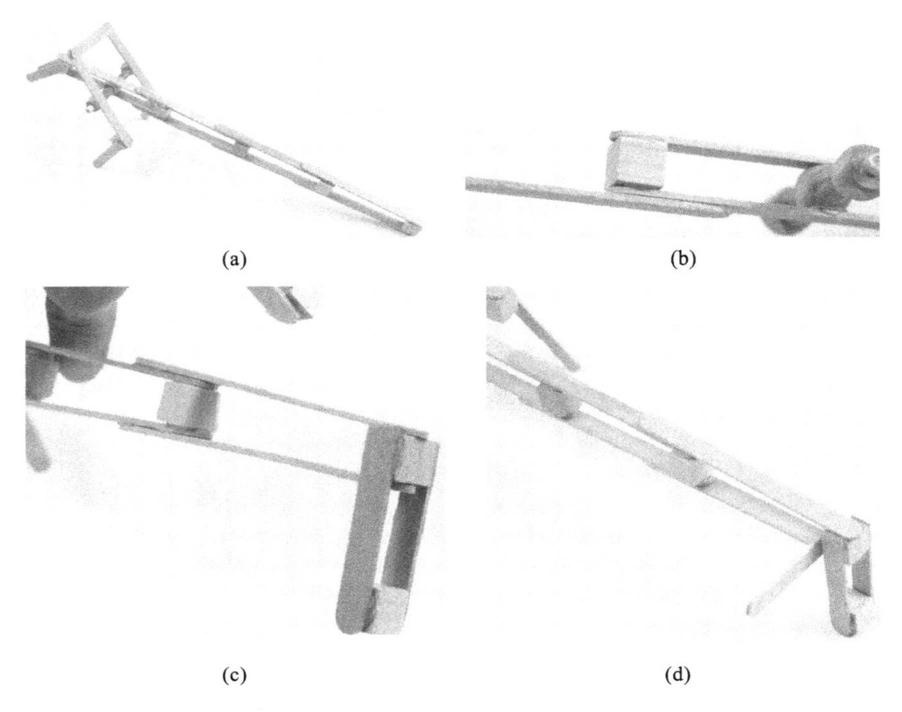

(a) (b) (c) (d)

Figure 5-4 Make the Arm

and cut off the excess. This can be tricky to do by oneself, so encourage yourself to help each other during this step.

Loosely tie a cable tie to hold the trigger in place. And finally, attach a rubber band to each of the outside fingers to the arm using a hitch knot. These will automatically open the hand when the user is not squeezing the trigger.

6. It's Alive!

Operation is simple. Hook your thumb over the thumb rest and place the handle at the base of your thumb. Wrap your fingers around the trigger and try it out! Try to pick up some everyday objects. What kind of things is the hand good at picking up? Where could it improve? How would you modify it? Ask yourself these questions.

The placement of the trigger is important for easy operation. You may need to adjust the trigger by as little as 1/2-inch to achieve optimal range of motion. Avoid using hot glue to secure the trigger since it is more difficult to adjust.

7. Tips and Troubleshooting①

This project offers a lot of opportunities to customize and redesign. Give

---

① troubleshooting 解决纷争；发现并修理故障

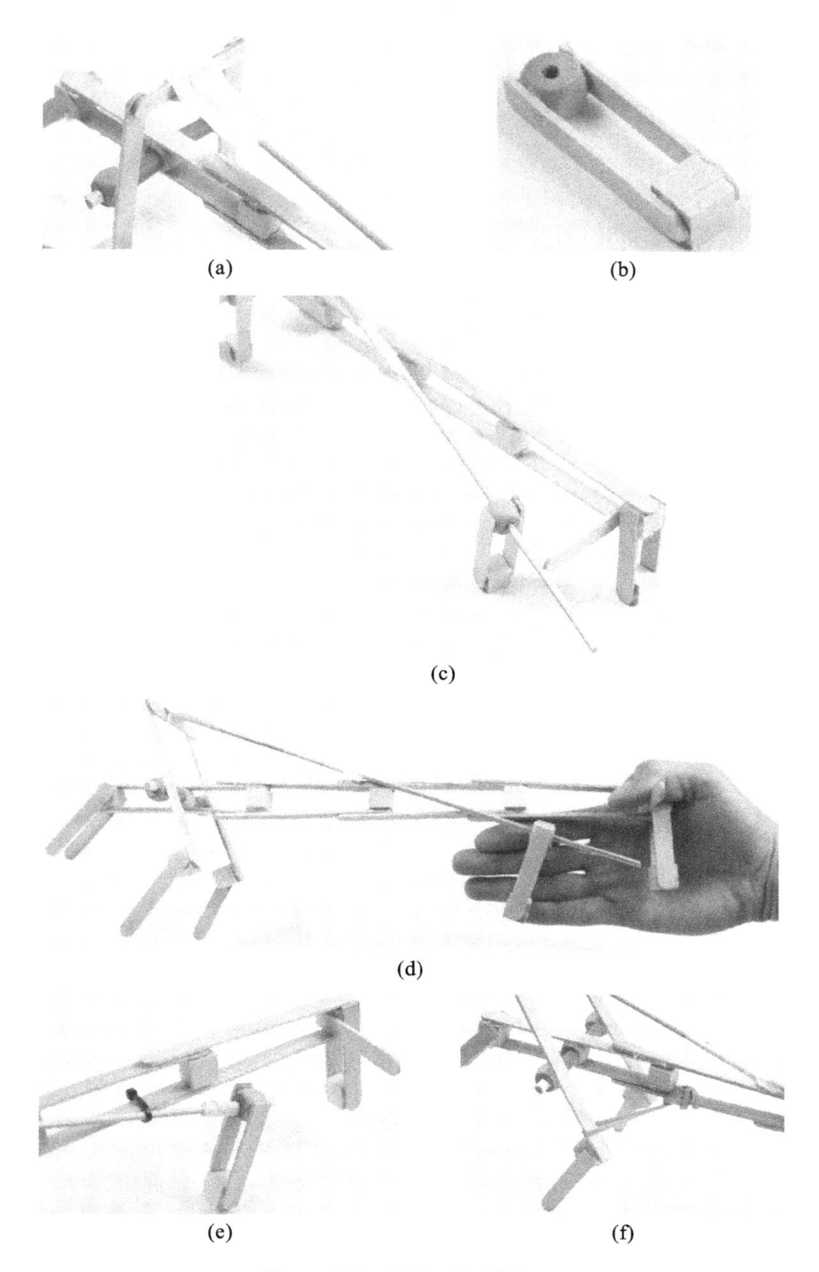

(a) (b)

(c)

(d)

(e) (f)

Figure 5-5 Make the Trigger

yourself at least two hours to: receive instruction, build the hand, and then redesign it while participating in the challenge.

Conducting the challenge is easy. Open a full box of straws and take turns trying to pick up as many as possible and move them into a separate container. Any straws that are dropped in transit don't count. Count the straws and keep a record. Straws will fall out and make a mess, you must be responsible for any straws that are dropped during your attempt.

This project has many steps and some small but important details. Having an example or two to refer to will help tremendously.

• If you have a small hand, then it will be more difficult to get a good range of motion. There are two ways to solve this. One is to lower the skewer that is attached to the back of the hand. This will bring it closer to the hinge. Smaller movements near the center of the hinge have a greater effect. Another solution is to make cable not tie the trigger to the arm, and instead use two hands to operate it: one hand holds the arm, and the other hand operates the trigger.

• If you want to make enormous mechanical hands and super-long arms, please remember that ultra-large-scale① designs are more susceptible to② breaking under its own weight, and the sheer amount of time required to build it leaves little time for testing and redesign.

• If the hand is difficult to open, try manually stretching the rubber bands to gain more slack.

5.2 Student Writing Guided-Lab Report

5.2.1 Before You Begin

The severity of any task is lessened when you take a moment to understand the purpose of your work. Before you begin writing, establish the issues you are going to address, who you are going to address them to, and why you need to do it.

1. Definition of a Lab Report

A lab report is a detailed account of an experiment, its methods, results, and conclusions which answer a question.

2. Define Your Discovery Question

Be sure to write down one or two primary "big picture" questions; your report addresses will be the focal point as you write your report. An example is as follows.

---

① ultra-large-scale 超大型，超大规模

② susceptible to 易受……影响的；对……敏感的

Discovery Question: What size electric heating element is installed in a given water heater?

3. Audience & Purpose

The description of audience and purpose is shown in Table 5-1.

Table 5-1 Audience and Purpose

Audience	• Engineers (Peers)	• Engineers interested in similar work will base their experiments on yours
	• Supervisors	• Supervisors want to know about the work you have done
	• TA	• The grader is also your audience
Purpose	• To Inform	• People want to know what you have done
	• To Persuade	• Raw data does not support itself; you must convince your audience it is correct

4. Why Write Well?

Recent surveys of Mechanical Engineering Faculty have shown that students need to be able to present their experimental results in an understandable way.

"Students do not understand how to sell their work/results. They have difficulty understanding what needs to be explained to the audience and what does not. They assume the audience knows what they know."—ME Faculty, 2007 WEC Survey.

5. Lab Report Elements

A report is created using the following characteristics.

Self-Supporting Document

This document can stand on its own. You are presenting enough information for the reader to understand the basis of your arguments. Other documents may be referenced for further investigation by the reader, such as a lab manual or journal article.

Name, Title, Page Number & Date

This document requires name, title, page number, and date. These are essential elements of formatting. Place your name or title with the page number in the header.

Standard Formatting

This document follows standard academic formatting guidelines. These include 12pt Font, 2. 5 cm margins, and headings which subdivide the information into manageable sections, with one heading per page minimum. Your instructor may have more stringent requirements.

Graphic Numbering

This document uses visuals. Each graphic, such as: figures, tables, pictures, equations, etc, is labeled and numbered sequentially. Word will manage this task for you—search Help for Captions and Cross-references.

IMRD Format

This document follows the IMRD traditional report writing standard. It contains the following sections in this order: **I**ntroduction, **M**ethods, **R**esults, and **D**iscussion. Introduction provides background and the question addressed, methods describes how that question was answered, results show the resulting data from the experiment and discussion is the author's interpretation of those results. Often results and discussion are combined.

Active Voice

This document encourages active voice. In active voice, the subject of a sentence is doing the action, such as, "I performed the experiment." This is different from the passive voice where the subject is receiving the action, such as, "the experiment was performed." Active voice adds clarity. It is becoming widely used, but you should still check with your instructor for their preference.

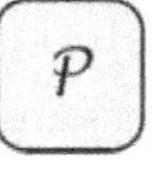

Persuasive

This document is trying to make the audience believe something.

6. Tense

Technical writing varies its tense depending on what you are discussing. Tense should be consistent for each section you write.

Past Tense

This document uses past tense. As a rule of thumb, past tense is used to describe work you did over the course of the report timeline.

Present Tense

This report uses present tense. As a rule of thumb, present tense is used to describe knowledge and facts that were known before you started.

Be consistent. Write a section in a consistent tense.

7. Why This Format?

In its early days, technical communication—the ability to communicate logic from one individual to another as efficiently as possible was not well developed.

Over the course of several hundred years, the standard IMRD format of the scientific paper was adopted as a standard. By the 1970s, nearly all academic journals required this standard for scientific experimental reporting. The basic outline is shown in Figure 5-6.

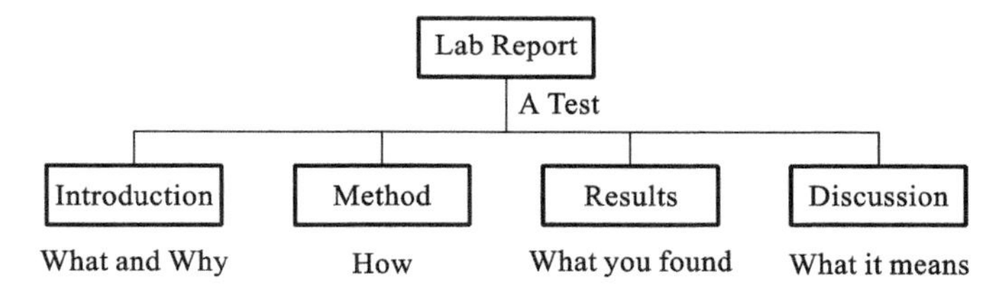

Figure 5-6 The basic Outline of the Standard IMRD Format

The report revolves around the solving of a specific question, described in the introduction and answered in the discussion.

5.2.2 How to Write a Lab Report

Table 5-2 describes the main content of a lab report.

Table 5-2 Main content of a lab report

Report Sections			Explanation
1.1	Title Page		
1.2	Abstract		
1.3	Table of Contents		
1.4	Introduction	• Background / Theory • Purpose • Governing Equations • Discovery Question (DQ)	In this section, you describe what you are trying to find and why. Background and motivation are used to provide readers with a reason to read the report

Continued Table 5-2

Report Sections			Explanation
1.5	Methods	• Experiment Overview • Apparatus • Equipment Table • Procedures	In this section, you explain how question addressed is answered. Clearly explain your work so it could be repeated
1.6	Results	• Narrate (like a story) • Tables and Graphs • Equations in Variable Forms • Uncertainties • Units • Indicate Final Results	In this section, you present the results of your experiment. Tables, graphs, and equations are used to summarize the results. Link equations and visuals together with narrative, like a story. Remember your audience
1.7	Discussion	• Answer DQ • Theoretical Comparison • Explanation of anomalies / Error • Conclusion / Summary • Future Work	In this section, you explain and interpret your results. Insert your opinion, backed by results. Discuss issues you had and how this could be corrected in the future. The conclusion is a summary of your results and discussion
1.8	References		
1.9	Appendices-raw data, sample calculations, lab notebook, etc		

1. Show Me!

The following sections show an example of lab report sections which have been annotated. In each section, **the black frame** indicates the required components and **the white frame** are suggestions to successfully write those parts.

1) Title Page

The title page contains a descriptive title of the document, the author's name, the affiliation, and the date. List any people who performed the work with you. This title page must conform to established standards of your class. Ask your instructor for his or her preference.

2) Abstract

The abstract's purpose is to summarize the information contained in the report for someone who doesn't have time or resources to read it, as shown in Figure 5-7. Its inclusion as a report "section" is slightly misleading. In many ways, the abstract is a document all on its own; it includes all the same parts of your report and its major findings.

Quantitative results and their uncertainties should be included when possible. It must contain parts from each major section of your report. Sometimes this is the only thing someone will read about your report. It should be no more than 400 words. This is not a "teaser".

You might be tempted to[1] write this first, as it is appears first chronologically[2] in the report; however, because the abstract is a summary of the entire report, you should write it last. This is the time when you are most familiar with the report and its major findings.

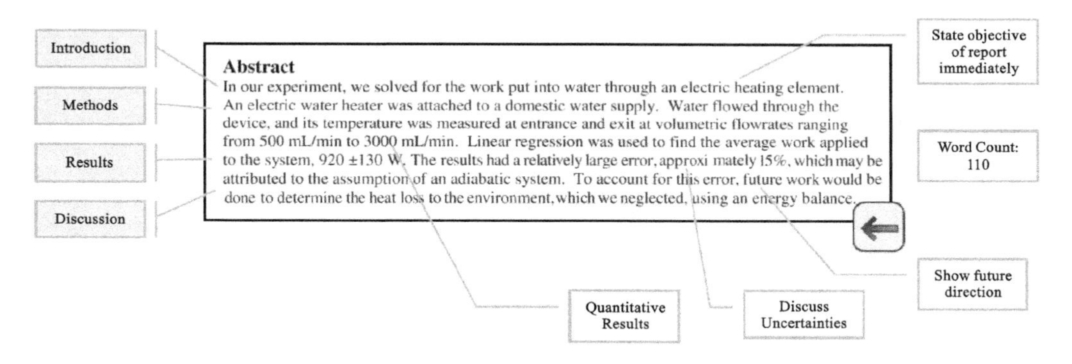

Figure 5-7 Abstract

3) Table of Contents

The table of contents' purpose is to allow the reader to find information easily. It also informs the reader about the report's organization. List page numbers with descriptive titles for the sections. This should be its own page.

4) Introduction

This explains what and why you are doing the experiment. It should show necessity of the experiment through theory and past work, as shown in Figure 5-8.

Explicitly stating the report question in the text of the introduction will help you keep the report in focus. As you continue writing, keep this question in mind—this is why you are making this report.

At this point, also notice that you haven't said anything about your experiment.

5) Methods

This section explains how the report question above was answered. After

---

① be tempted to 忍不住做某事，受诱惑做某事

② chronologically [ˌkrɒnəˈlɒdʒɪkli] *adv.* 按年月顺序排列的

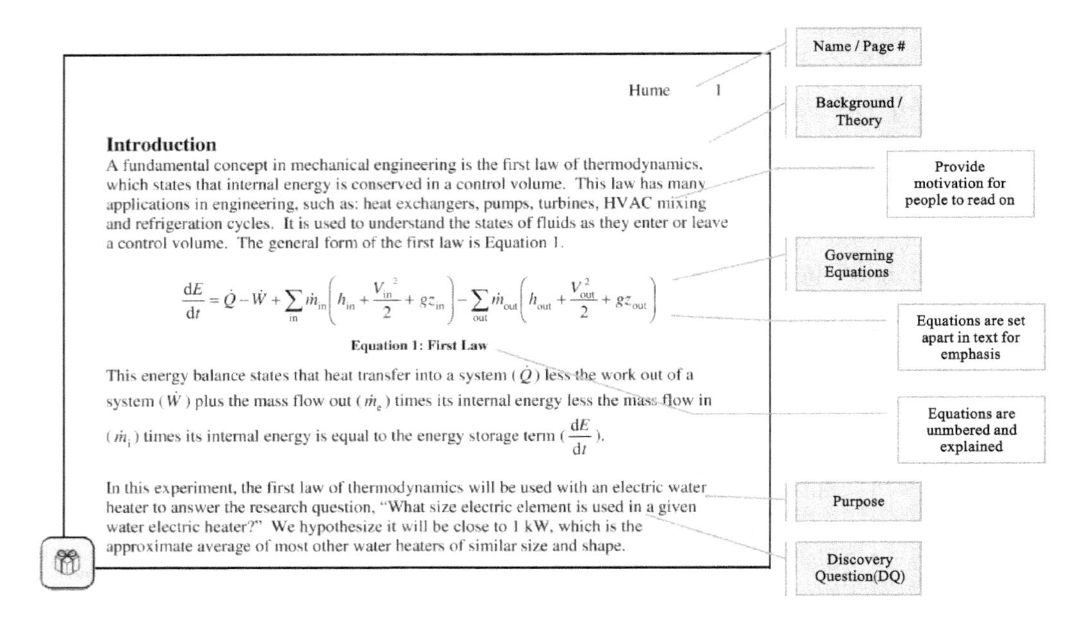

Figure 5-8 Introduction

reading this section, the reader should be able to completely reproduce the experiment to verify the results, as shown in Figure 5-9.

Notice the figure narration scheme so far. The report is a story of visuals linked together with text.

6) Results

It describes the results you achieved in your experiment, as shown in Figure 5-10.

So far, you have only presented your data. You haven't described what it means. That comes in the next section.

7) Discussion

In this section, the results are interpreted. Describe the reason why you think the data turned out like it did. Insert your scientific opinion in this section ,as shown in Figure 5-11.

The report is now fully described.

8) Conclusion / Summary

This section is a summary of the results and discussion from the report. It is still discussion, where you insert your opinion of the results. Report the key findings of the report here. It is much like the results and discussion sections of the abstract. Directly answer the report question here. Do not be vague, as shown in Figure 5-12.

Hume 2

Method

Tap water was passed through the electric water heater at flow rates ranging from 50 mL/min to 300 mL/min. The input and output temperatures were recorded. The information was then used with the first law to approximate the work by the heating element performed on the system.

Experiment Overview

A high level description of the experiment instantly informs the audience

"Describe the forest before you describe the trees"

Apparatus

The experimental apparatus includes a plastic container, the heater, measuring 6" x 6" x 12". One half inch ports were located at the upstream and downstream sides of the heater and labeled 'in' and 'out'. An electric power cord is attached to the top side of the heater and supplies 120Vac to the heater inside. The input to the heater was connected to a domestic water supply, which provided 46.1 ± 0.4 °F water once at steady state. The output from the water heater was run through a Dwyer flowmeter, which measured and controlled flowrate, and then exited into a sink. Temperature readings were made at the inlet and exit ports of the heater with T thermocouples and an Omega temperature indicator. The experimental setup is shown in Figure 1.

Figure 1: Apparatus

A detailed list of the equipment used in this experiment and their uncertainties are shown in Table 1.

Table 1: Equipment

Equipment	Uncertainty
Electric Heater	n/a
Dwyer RMA-14 Rotameter	+/− 50 mL/min
Omega T Thermocouples	+/− 1.8 °F
Omega Temp Readout	n/a

Procedure

The tap water entering the system was varied between 500 ml/min and 3000 ml/min, in 500 ml/min increments. At each flowrate, the inlet and outlet temperatures were measured ten times when the system appeared to reach steady state.

Apparatus Description

"Walk through" the apparatus description

Apparatus Sketch

Enable reader to visualize the experiment

Use descriptive annotations

Figure Numbers & Captions

Every equation, table, or figure is discussed in the text

Equipment Table

Manufacturer, model number, serial number, uncertainties

Procedure

NO bulleted statements

Step-by-Step cookbook instructions

Figure 5-9 Methods

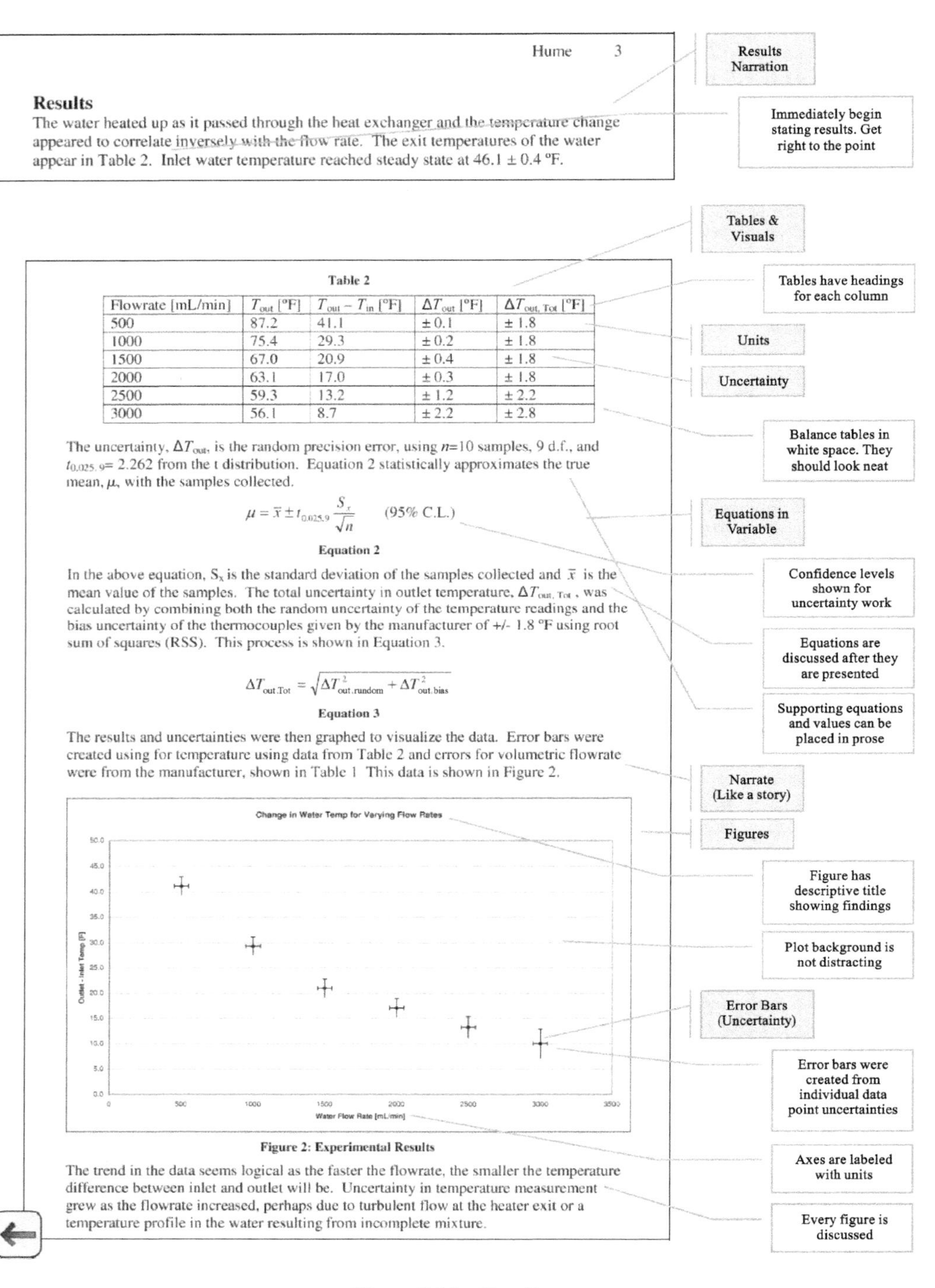

Hume 3

Results

The water heated up as it passed through the heat exchanger and the temperature change appeared to correlate inversely with the flow rate. The exit temperatures of the water appear in Table 2. Inlet water temperature reached steady state at 46.1 ± 0.4 °F.

Table 2

Flowrate [mL/min]	T_{out} [°F]	$T_{out}-T_{in}$ [°F]	ΔT_{out} [°F]	$\Delta T_{out,\,Tot}$ [°F]
500	87.2	41.1	± 0.1	± 1.8
1000	75.4	29.3	± 0.2	± 1.8
1500	67.0	20.9	± 0.4	± 1.8
2000	63.1	17.0	± 0.3	± 1.8
2500	59.3	13.2	± 1.2	± 2.2
3000	56.1	8.7	± 2.2	± 2.8

The uncertainty, ΔT_{out}, is the random precision error, using n=10 samples, 9 d.f., and $t_{0.025,9}$= 2.262 from the t distribution. Equation 2 statistically approximates the true mean, μ, with the samples collected.

$$\mu = \bar{x} \pm t_{0.025,9}\frac{S_x}{\sqrt{n}} \quad \text{(95\% C.L.)}$$

Equation 2

In the above equation, S_x is the standard deviation of the samples collected and $\bar{x}$ is the mean value of the samples. The total uncertainty in outlet temperature, $\Delta T_{out,\,Tot}$, was calculated by combining both the random uncertainty of the temperature readings and the bias uncertainty of the thermocouples given by the manufacturer of +/- 1.8 °F using root sum of squares (RSS). This process is shown in Equation 3.

$$\Delta T_{out.Tot} = \sqrt{\Delta T^2_{out.random} + \Delta T^2_{out.bias}}$$

Equation 3

The results and uncertainties were then graphed to visualize the data. Error bars were created using for temperature using data from Table 2 and errors for volumetric flowrate were from the manufacturer, shown in Table 1 This data is shown in Figure 2.

Figure 2: Experimental Results

The trend in the data seems logical as the faster the flowrate, the smaller the temperature difference between inlet and outlet will be. Uncertainty in temperature measurement grew as the flowrate increased, perhaps due to turbulent flow at the heater exit or a temperature profile in the water resulting from incomplete mixture.

Figure 5-10 Results

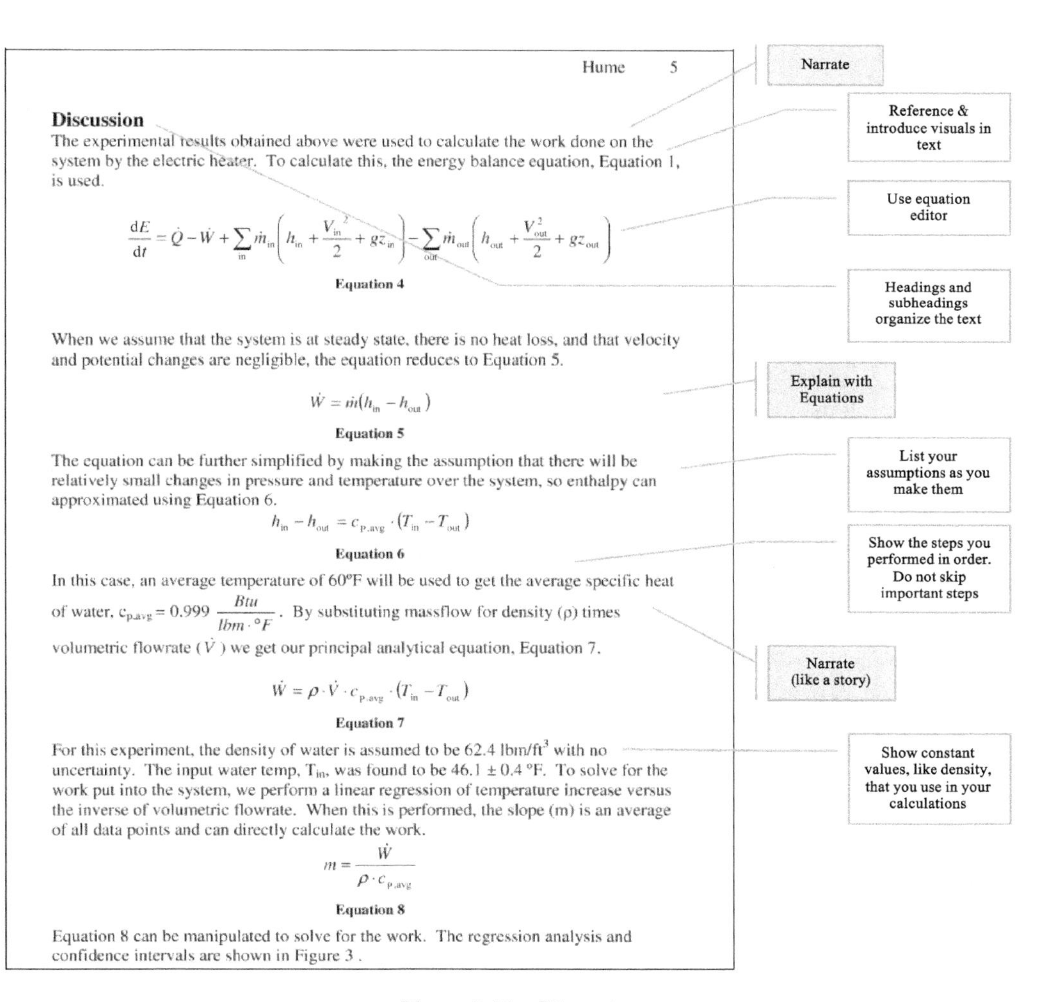

Hume 5

Discussion

The experimental results obtained above were used to calculate the work done on the system by the electric heater. To calculate this, the energy balance equation, Equation 1, is used.

$$\frac{dE}{dt}=\dot{Q}-\dot{W}+\sum_{in}\dot{m}_{in}\left(h_{in}+\frac{V_{in}^2}{2}+gz_{in}\right)-\sum_{out}\dot{m}_{out}\left(h_{out}+\frac{V_{out}^2}{2}+gz_{out}\right)$$

Equation 4

When we assume that the system is at steady state, there is no heat loss, and that velocity and potential changes are negligible, the equation reduces to Equation 5.

$$\dot{W}=\dot{m}(h_{in}-h_{out})$$

Equation 5

The equation can be further simplified by making the assumption that there will be relatively small changes in pressure and temperature over the system, so enthalpy can approximated using Equation 6.

$$h_{in}-h_{out}=c_{p,avg}\cdot(T_{in}-T_{out})$$

Equation 6

In this case, an average temperature of 60°F will be used to get the average specific heat of water, $c_{p,avg}=0.999\ \frac{Btu}{lbm\cdot{}^{\circ}F}$. By substituting massflow for density (ρ) times volumetric flowrate ($\dot{V}$) we get our principal analytical equation, Equation 7.

$$\dot{W}=\rho\cdot\dot{V}\cdot c_{p,avg}\cdot(T_{in}-T_{out})$$

Equation 7

For this experiment, the density of water is assumed to be 62.4 lbm/ft^3 with no uncertainty. The input water temp, T_{in}, was found to be 46.1 ± 0.4 °F. To solve for the work put into the system, we perform a linear regression of temperature increase versus the inverse of volumetric flowrate. When this is performed, the slope (m) is an average of all data points and can directly calculate the work.

$$m=\frac{\dot{W}}{\rho\cdot c_{p,avg}}$$

Equation 8

Equation 8 can be manipulated to solve for the work. The regression analysis and confidence intervals are shown in Figure 3 .

Figure 5-11 Discussion

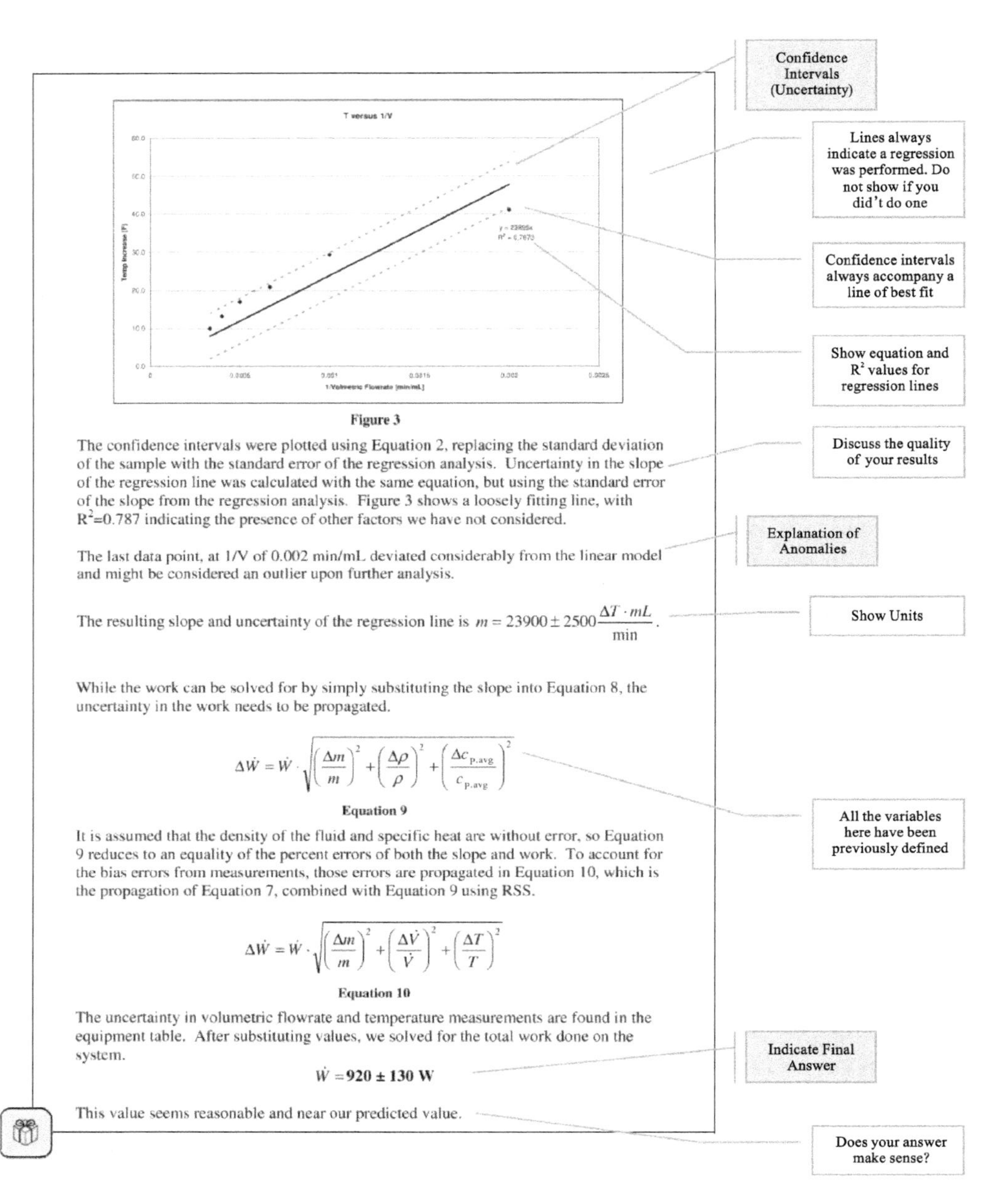

Figure 3

The confidence intervals were plotted using Equation 2, replacing the standard deviation of the sample with the standard error of the regression analysis. Uncertainty in the slope of the regression line was calculated with the same equation, but using the standard error of the slope from the regression analysis. Figure 3 shows a loosely fitting line, with R^2=0.787 indicating the presence of other factors we have not considered.

The last data point, at 1/V of 0.002 min/mL deviated considerably from the linear model and might be considered an outlier upon further analysis.

The resulting slope and uncertainty of the regression line is $m = 23900 \pm 2500 \frac{\Delta T \cdot mL}{\min}$.

While the work can be solved for by simply substituting the slope into Equation 8, the uncertainty in the work needs to be propagated.

$$\Delta \dot{W} = \dot{W} \cdot \sqrt{\left(\frac{\Delta m}{m}\right)^2 + \left(\frac{\Delta \rho}{\rho}\right)^2 + \left(\frac{\Delta c_{p,avg}}{c_{p,avg}}\right)^2}$$

Equation 9

It is assumed that the density of the fluid and specific heat are without error, so Equation 9 reduces to an equality of the percent errors of both the slope and work. To account for the bias errors from measurements, those errors are propagated in Equation 10, which is the propagation of Equation 7, combined with Equation 9 using RSS.

$$\Delta \dot{W} = \dot{W} \cdot \sqrt{\left(\frac{\Delta m}{m}\right)^2 + \left(\frac{\Delta \dot{V}}{\dot{V}}\right)^2 + \left(\frac{\Delta T}{T}\right)^2}$$

Equation 10

The uncertainty in volumetric flowrate and temperature measurements are found in the equipment table. After substituting values, we solved for the total work done on the system.

$$\dot{W} = 920 \pm 130 \text{ W}$$

This value seems reasonable and near our predicted value.

Continued Figure 5-11

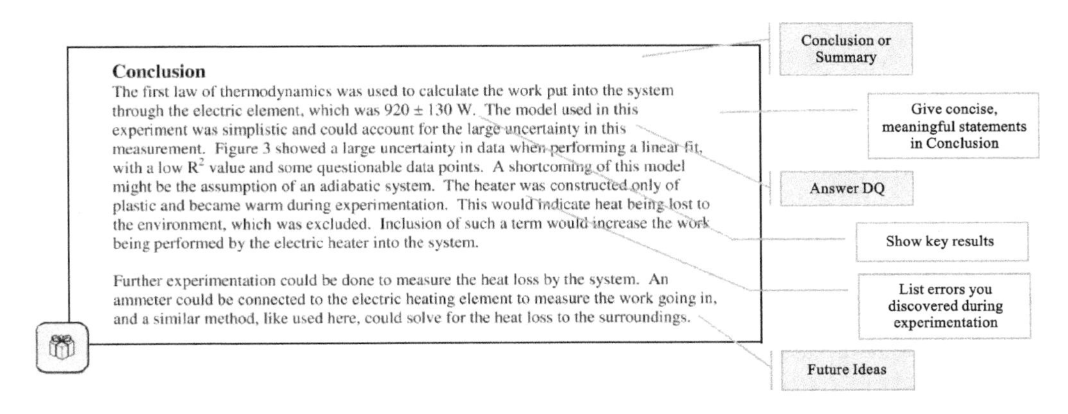

Conclusion

The first law of thermodynamics was used to calculate the work put into the system through the electric element, which was 920 ± 130 W. The model used in this experiment was simplistic and could account for the large uncertainty in this measurement. Figure 3 showed a large uncertainty in data when performing a linear fit, with a low R^2 value and some questionable data points. A shortcoming of this model might be the assumption of an adiabatic system. The heater was constructed only of plastic and became warm during experimentation. This would indicate heat being lost to the environment, which was excluded. Inclusion of such a term would increase the work being performed by the electric heater into the system.

Further experimentation could be done to measure the heat loss by the system. An ammeter could be connected to the electric heating element to measure the work going in, and a similar method, like used here, could solve for the heat loss to the surroundings.

Figure 5-12 Conclusion

9) References

The reference section shows where you got information that was not your own, as shown in Figure 5-13.

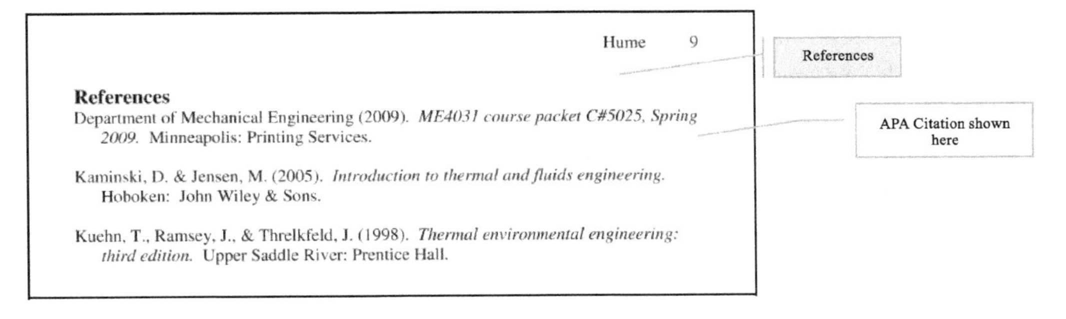

Hume 9

References

Department of Mechanical Engineering (2009). *ME4031 course packet C#5025, Spring 2009.* Minneapolis: Printing Services.

Kaminski, D. & Jensen, M. (2005). *Introduction to thermal and fluids engineering.* Hoboken: John Wiley & Sons.

Kuehn, T., Ramsey, J., & Threlkfeld, J. (1998). *Thermal environmental engineering: third edition.* Upper Saddle River: Prentice Hall.

Figure 5-13 References

There are many citation styles you can use, such as ASME, CMS, APA, etc. Consult a citation manual for assistance. *A Pocket Style Manual* by Diana Hacker is a good start.

10) Appendices

The appendix should contain information that is required, but would be distracting from the normal flow of the report. This might be raw data, lab notebook pages, regression summaries, or sample calculations.

2. Process Tips

• It is more important you are clear and direct than to follow formatting rules.

• The report organization doesn't follow the way to write it, you need to think about the best order to write it. To help, write a report in the following order: Methods, Results, Discussion, Intro, and Abstract.

- Use visuals. Engineering is more than prose writing.
- Be concise. Extra words actually detract from meaning.
- Think of a report as a big string of visuals, linked together by narrative sentences.
- Graphs, figures, tables, and equations are all worthy of their own line.
- Avoid showing actual calculations in the body of the report—they are difficult to understand. Keep everything in variable format, and show numerical calculations in the appendix.
- Some instructors require more rigorously① formatted reports. Check with them if you have any questions.

3. Assessment Criteria

Please follow the lab report writing checklist when you write a lab report.

- **Cover Page**
- **Abstract** gives a quick, complete summary of the experiment and its conclusions. Less than 400 words.
- **Table of Contents**
- **Introduction** provides background and theory for the experiment; shows what the experiment will find and why it is needed. States DQ.
- **Method** gives a complete description of the apparatus, equipment, and procedure which was followed in the experiment.
- **Results** describe the data obtained when the method was performed; they show uncertainties.
- **Discussion** is your interpretation of the results and describes them like a story; it answers DQ.
- **References**
- **Appendix**

① rigorously [ˈrɪɡərəsli] *adv.* 严厉地；残酷地

Chapter 6 Different Ways to Use Professional English

6.1 How to Write an Abstract for a Scientific Paper

If you're preparing a research paper or grant proposal, you'll need to know how to write an abstract. Here's a look at what an abstract is and how to write one.

What is an Abstract?

An abstract is a concise summary of an experiment or research project. It should be brief—typically under 200 words. The purpose of the abstract is to summarize the research paper by stating the purpose of the research, the experimental method, the findings, and the conclusions.

How to Write an Abstract

The format you'll use for the abstract depends on its purpose. If you're writing for a specific publication or a class assignment, you'll probably need to follow specific guidelines. If there isn't a required format, you'll need to choose from one of two possible types of abstracts.

1) Informational Abstracts

An informational abstract is a type of abstract used to communicate an experiment or a lab report.

• An informational abstract is like a mini-paper. Its length ranges from a paragraph to 1-2 pages, depending on the scope of the report. Aim for less than 10% the length of the full report.

• Summarize all aspects of the report, including purpose, method, results, conclusions, and recommendations. There are no graphs, charts, tables, or images in an abstract. Similarly, an abstract does not include bibliographies[①] or references.

---

① bibliography [ˌbɪbliˈɒgrəfi] *n.* 参考书目；文献目录

· Highlight important discoveries or anomalies[①]. It's okay if the experiment did not go as planned and it's necessary to state the outcome in the abstract.

Here is a good format to follow, in order, when writing an informational abstract. Each section is a sentence or two long.

(1) **Motivation or Purpose**: State why the subject is important or why anyone should care about the experiment and its results.

(2) **Problem**: State the hypothesis[②] of the experiment or describe the problem you are trying to solve.

(3) **Method**: How did you test the hypothesis or try to solve the problem?

(4) **Results**: What was the outcome of the study? Did you support or reject a hypothesis? Did you solve a problem? How close were the results to what you expected? State specific numbers of your study.

(5) **Conclusions**: What is the significance of your findings? Do the results lead to an increase in knowledge, solution that may be applied to other problems, etc. ?

2) Descriptive Abstracts

A descriptive abstract is an extremely brief description of the contents of a report. Its purpose is to tell the reader what to expect from the full paper.

· A descriptive abstract is very short, typically less than 100 words.

· It tells the reader what the report contains, but doesn't go into details.

· It briefly summarizes the purpose and experimental method, but not the results or conclusions. Basically, it says why and how the study was made, but doesn't go into findings.

Tips for Writing a Good Abstract

· Write the paper before writing the abstract. You might be tempted to start with the abstract, since it comes between the title page and the content, but it's much easier to summarize a paper or report after it has been completed.

· Write in the third person. Replace phrases like "I found" or "we examined" with phrases like "it was determined" or "this paper provides" or "the investigators found".

· Write the abstract and then pare it down to meet the word limit. In some cases, a long abstract will result in automatic rejection of publication or a grade!

· Think of keywords and phrases a person looking for your work might use or enter into a search engine. Include those words in your abstract. Even if the paper

---

① anomaly [əˈnɒməli] *n*. 异常现象，反常现象

② hypothesis [haɪˈpɒθəsɪs] *n*. 假设

won't be published, this is a good habit to develop.

- All information in the abstract must be covered in the body of the paper. Don't put a fact in the abstract that isn't described in the report.
- Proof-read the abstract for typos, spelling mistakes, and punctuation errors.

6.2 How to Deal With Difficult Sentences

It is common for a Chinese to misunderstand sentences and paragraphs even thought about all the vocabularies are known. The difficulty is probably related to sentences that are complex or long with science-based types of words. Here are some suggestions on what to do under these situations.

In order to understand complex sentences in science, you could first identify the "**headword**" and work backwards①, because it is often the words coming after the headword that cause the greatest difficulty. What is a headword? It is a noun formed from a verb, modified by adjectives, nouns, or other words which may come before or after it. Thus in the sentence "*The outcome of all the meetings was a revised research methodology*②", the headword is "*outcome*". Look at the following long sentence:

*Predictions of the origins of viral*③ *strains based on measured changes in DNA will be most accurate if we examine only viruses that are mosquito*④ *borne.*

The headword is "*predictions*" because it is what the main message in the sentence is about. We could work backward by changing the noun back to a verb ("predict") and analyse⑤ which noun, functioning as subject and object, is associated with it. The subject of predictions in the example is "*viruses*", or more specifically, "*the origins of viral strains*". The rest of the sentence can now be determined: changes in DNA of viruses pertaining to mosquito vectors. Therefore the sentence is about using DNA to predict the origin of viruses carried by mosquitoes.

The following is an even longer sentence:

After an experiment which exposed the animals to various levels of

---

① backwards ['bækwədz] *adv.* 倒；向后；逆

② methodology [ˌmeθə'dɒlədʒi] *n.* 方法学，方法论

③ viral ['vaɪrəl] *adj.* 滤过性毒菌引起的；滤过性毒菌的

④ mosquito [mə'skiːtəʊ] *n.* 蚊子

⑤ analyse ['ænəlaɪz] *vt.* 分析；分解；细察

disturbance, recommendations were drafted by the government's conservation body for a buffer zone of 200 m to reduce disturbance by human recreational activities such as outboard-power boats.

In this example, "*recommendations*" is the headword from the verb "recommend" and consequently the subject (the one who recommends) is the government. The main point of the sentence is the government made recommendations to reduce human disturbance.

There may be one or more verbs in a complex or long sentence which could confuse the reader. The confusion is related to which nouns in the sentence should function as subject or object of the verb(s). In such situation, it is useful to identify the **main verb** in the sentence and ask what or who is doing the action. In the following example the main verb has been identified (highlighted):

*Logarithmic*① *transformation of one or both of the axes of a curvilinear*② *relationship will usually* ***result*** *in an adequate straightening.*

Once the main verb ("*result*") has been identified, you could ask what results in what. In this case it is logarithmic transformation that results in straightening (of the curvilinear relationship) and not curvilinear relationship that results in the straightening.

The sentence below contains the verbs "*cooling*", "*cause*", "*expel*", and "*contain*". The main verb in the sentence, however, is "*causes*".

Cooling the metal from higher temperatures to below its transition temperature ***causes*** *it to expel any magnetic flux which it might contain.*

Having identified the main verb, you could now ask further questions such as "What causes what to do what?" or "What did it cause?" The sentence clearly infers that cooling the metal is what causes the reaction, and the reaction is to expel magnetic flux.

Although the occurrence of several verbs could cause much confusion, the text becomes easier to read once the main verbs are identified. Here is another example:

Sometimes plant or animal tissues are ***broken down*** *by heat and pressure beneath the earth's surface,* ***rejecting*** *oxygen, hydrogen, and nitrogen and* ***leaving behind*** *a residue or elemental carbon which in many cases* ***reflects*** *the appearance of plant or animal.*

Upon identifying the main verbs, you could follow up questions beginning with

---

① logarithmic [ˌlɒgəˈrɪðmɪk] *adj.* 对数的

② curvilinear [ˌkɜːvɪˈlɪniə] *adj.* 曲线的；由曲线组成的(等于 curvilineal)

"What", "Which", "Why", or "How". For example:

What are broken down by heat and pressure? (Answer: plant or animal tissues)

Which elements are rejected? (Answer: oxygen, hydrogen, nitrogen)

What is left behind? (Answer: residue or elemental carbon)

How is elemental carbon formed? (Answer: breakdown of plant or animal tissues)

Why does the residue "*reflect*" the appearance of plant or animal? (Answer: because the residue comes from the plant or animal in the first place)

There could be more questions which may be asked. In all cases, as before, identify the main verbs and then simply look at the sentence to answer the questions posed. Often long sentences are joined together by words such as "and", "or", and "but". Such **joining words** can also cause problems to Chinese readers, who may be confused by what is exactly joined to what. The best way is to separate the phrase① preceding the joining word from the phrase immediately after the joining word. For example,

Because mercury does not have enough mass for its self-gravitation② to produce significant compressing of its core, we can only conclude that the planet's high density is due to the material that forms the innermost planet having ***a higher proportion of heavier elements such as iron*** *and* ***the corresponding depletion of lighter elements such as carbon③.***

The phrases immediately before and after the joining word "*and*" are highlighted. Now, working out the rest of the sentence is easier; the main point expressed is despite mercury lacking mass to produce compression of its core, the planet has a high density.

In the sentence below "*and*" separates the two island groups of populations to show that they are genetically differentiated:

The two groups of populations were well differentiated, with ***the outer island populations having allele④ A fixed in the group*** *and* ***the inner island population having low frequencies of allele A*** *but high frequencies of allele B.*

In the sentence below, "*or*" is used to join "*by removing some quantities of*

---

① phrase [freɪz] *n.* 短语；习语；措辞；[音]乐句

② gravitation [ˌgrævɪˈteɪʃn] *n.* 重力；万有引力；地心吸力

③ carbon [ˈkɑːbən] *n.* [化学] 碳；碳棒；复写纸

④ allele [əˈliːl] *n.* 等位基因

plant tissue" and "*by directly obtaining nutrients from the mesophyll*①" to show that either circumstance can affect the hosts.

*In contrast to the majority of phytophagou*② *insects that feed externally, gall-inducing insects may affect the inner organs of their hosts* ***by removing some quantities of plant tissue*** *or* ***by directly obtaining nutrients from the mesophyll*** *of leaves, stems, flowers, or roots.*

Notice the word "*by*" is repeated for the joining phrase. This is the correct usage. It is incorrect to omit the second "*by*" in this sentence (as "*by removing some quantities of plant tissue* ***or directly*** *obtaining nutrients from the mesophyll*") because the phrase immediately after "*or*" would not link up with "... *may affect the inner organs of their hosts*".

Notice that often a sentence on its own makes little sense without reference③ to other sentences in the text. A stand-alone sentence may give a message or make a statement, but to fully understand the message or statement you must appreciate how it fits in with other sentences. To understand the text, you must recognize how the sentences are connected in the text. Obvious ways of connecting sentences are through the use of connecting words such as "also" and "consequently", of reference words such as pronouns (it, these, we, etc.), of substitution words such as "therefore" and "such", and of words referring to the same thing such as synonyms or simply repeating the same word. Scientists also tend to omit (or modify) words that the reader is expected to understand because they were used in a previous sentence. Consider the paragraph below:

*The island wind charts for the year 1995-2005 are shown in Fig. 2. The year 2002 has a much higher share of easterly wind directions than the other years, where usually winds from the southwest clearly dominate. Similar meteorological*④ *trend was observed on the mainland (12 km east of the island). These data do not represent the regional wind flow. Therefore local wind direction may have caused the higher concentration of pollutants measured on the island in 2002.*

The second sentence (beginning with "*The year 2002...*") contains the phrase "*... than the* ***other*** *years*", in which "*other*" substitutes⑤ "1995-2005" as stated in

---

① mesophyll [ˈmesəfɪl] *n.* [植] 叶肉

② phytophagous [faɪˈtɒfəgəs] *adj.* (昆虫)食植物的；食叶类的

③ reference [ˈrefrəns] *n.* 参考，参照；涉及，提及；考书参目；介绍信；证明书

④ meteorological [ˌmiːtiərəˈlɒdʒɪkəl] *adj.* 气象的；气象学的

⑤ substitute [ˈsʌbstɪtjuːt] *vt.* 取代；代替

the preceding sentence. Similarly, it is difficult to understand the third sentence without knowing that the "*Similar meteorological trend*" refers to "*easterly wind*" mentioned in the previous sentence. Likewise, "*These data*" in the next sentence refers to the wind observations as stated in the previous sentences. The last sentence is closely connected to all previous sentences in the paragraph, and would not make much sense if it is read as a stand-alone sentence.

Successful reading is of course influenced by the student's general proficiency in English language, particularly English grammar. It is beyond the scope of this book to cover the fundamentals of English grammar, although **Chapter** 7 covers in some depth common grammatical mistakes committed by Chinese. You should, however, already be familiar with basic sentence structures so that you can make distinctions① when different sentence structures are used to convey the same meaning and when different sentence structures convey a difference in meaning. The meaning of the following two sentences is basically the same.

There is an increasing realization that domestic-cat monitoring schemes with conservation implications require practical involvement of the human community.

Monitoring of domestic cat movement for conservation purposes requires community involvement.

Whereas the following two sentences convey a different meaning (make sure you know the differences).

Studies commonly find that smaller wooded patches support fewer species than larger patches, presumably② because larger patches reflect higher resource levels.

Studies show that an increase in species with patch size can either be due to habitat diversity in the larger patches or to area size per se.

Know how to use verb tenses, even though these are commonly misused by Chinese owing to cultural discrepancy. In English a change in the tense could make the meaning in a statement quite different. A sentence such as "*The Government has developed a plan to mitigate③ flooding of the river*" may not mean the same as "*The Government develops a plan to mitigate flooding of the river*". The first sentence states that a plan is in place already, but the second sentence tends to imply a planning stage only.

---

① distinction [dɪ'stɪŋkʃn] *n.* 区别；差别；特性；荣誉，勋章

② presumably [prɪ'zjuːməbli] *adv.* 大概；推测起来；可假定

③ mitigate ['mɪtɪgeɪt] *vt.* 使缓和，使减轻

Have a reasonable understanding of how sentences are related through the use of words that do not have meaning themselves but refer to something else for their meanings; examples are "he", "these", "this", "those", "that", "here", "there", "them". Consider the following paragraph.

The variables *were initially examined by bivariate*[①] *correlations to determine significant pair-comparisons.* ***Those*** *that were expressed as percentages were transformed with arcsine* $\sqrt{X}$. ***They*** *were then subjected to principal component analysis.*

To understand the text, you need to know what the words "*those*" and "*they*" refer to in the text. In the example, "*those*" refers to "*the variables*" and "*they*" refers to the transformed variables. Note that "*those*" does not refer to "*pair-comparisons*" and "*they*" does not refer to "*percentages*".

Understand relationships between clauses[②] that employ conjunctions. Thus "*However*" below means the opposite to the ideas that have stated before.

The question of will sustainable technology become competitive with the existing technology is yet to be answered. ***However***, *we know that pricing of many raw materials deriving from non-renewable resources is increasing.*

Other conjunctions are "for example", "although" and "furthermore", etc.

Be aware that scientists often avoid repeating a word by substituting the word with another and by omitting certain words in a phrase. For example:

The number of ***visits by tourists*** *reached a peak in the summer months. However,* ***many*** *were restricted to the edge of the nature reserve. Only* ***a few*** *were allowed into the reserve proper to minimize disturbance to the wildlife. In* ***all cases***, *only trampling of soil remains an issue.*

In this example, "*visits by tourists*" has been omitted or substituted by the use of the words "*many*" (which means "many visits by tourists"), "*a few*" (which means "a few visits by tourists"), and "*all cases*" (which means "all cases of visits by tourists").

In summary, scientific paper often contains complex ideas that are not always easy to understand. You read to gain knowledge, and the more you read the more knowledge you gain. Knowledge is a building process: your capacity to understand and learn will increase as you become more familiar with a particular subject. Also the more you read, the better you will understand difficult sentences, as long as

① bivariate [baɪ'værɪɪt] *adj*. 二变量的

② clause [klɔːz] *n*. 条款;[计] 子句

you ensure that you understood and learnt from difficult sentences you have encountered previously.

6.3 How to Write Personal Résumé

There are three basic guidelines for effective résumé writing, including:

(1) discovering what kind of jobs fit your interests and skills;

(2) locating job opportunities that match your interests and skills; and

(3) tailoring a résumé to fit the qualifications of a particular job.

The place to start is to think about what kind of work you like to do, what kind of work you do well, and what kind of jobs fit your interests and skills best. Once you've narrowed your focus to a specific set of possible jobs, then you must locate actual job openings and examine the descriptions of those openings. Finally, you must tailor your résumé to fit the qualifications a potential employer wants and needs.

6.3.1 Self-analysis

Before you start thinking about what kind of jobs you may want to apply for, you need to figure out the following things.

- what you do well
- what skills you possess, and
- what kind of work you enjoy

Writing a self-analysis paper before you begin thinking about specific jobs will help clarify your previous successes, established skills, and aspirations for the future.

Self-analysis is a tool that many employers use as part of their overall assessment strategies. While older models of assessment were limited to a "top-down" approach where only the boss got to evaluate his or her workers, more and more employers want assessment to be a two-way street. In other words, employers often ask their employees to write self-analyses along with the boss' own work evaluations. Then the employer sits down with the employee, and they discuss each other's perceptions about the employee's work performance. This new model of assessment helps clear up misunderstandings between employer and employee and offers the employee an opportunity to think about his or her own strengths and weaknesses. Writing a self-analysis paper not only helps you to create a better résumé; it also helps you prepare for future assessment meetings with a potential employer.

In addition, because a self-analysis demands that you take the time to write down your occupational desires and work strengths, you will be better prepared in an interview to talk about these aspects of your job qualifications. Whenever we write something down about ourselves, we tend to remember what we wrote and are better able to explain it to other people.

The following self-analysis asks various questions about what kind of work you do best and what sort of jobs you want most, as shown in Table 6-1. Try to answer the questions as honestly as you can, since they will provide you with the fundamental information you need to begin your job search and résumé.

Table 6-1 Self-analysis

What specific job do you want?
List three duties of a person with this job title.
Describe the kind of companies you would like to work for (size, location, type of product/service provided). If you have a specific company in mind, use that company.

Continued Table 6-1

What work-related activities do you enjoy most and why?
What work-related activities do you enjoy least and why?
What was the most successful thing you ever did and why did it successfully?
What was the least successful thing you ever did and why did it unsuccessfully?
What traits of yours accounted for the success and failure noted in the previous questions?

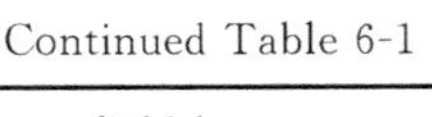

Continued Table 6-1

Name three of your work-related skills or experiences that not everyone in your field has.
List three of your long-range career goals.
Fill in the blank in the following sentence with one or more words from the list given (or one of your own). I am a wonderful... (communicator, conversationalist, counselor, creator, detail person, designer, organizer, salesperson, teacher, team leader)
Circle five words from the following list that describe you. accurate, assertive, creative, dependable, diligent, disciplined, effective, efficient, experienced, flexible, hardworking, intelligent, likable, mature, personable, polite, productive, reliable, resourceful, self-confident, self-motivated, well-groomed, well-trained

Continued Table 6-1

Describe yourself to a potential employer in about fifty (50) words.

6.3.2 Interests and Skills

What skills do you already possess, and how do those skills connect with your job interests? One way to answer these questions is to put your skills in three different contexts.

- Skills with People
- Skills with Things
- Skills with Information or Data

For example, if you are a good communicator and enjoy consulting with others, you have strong skills with people. On the other hand, if you are good with your hands and enjoy building things or operating equipment, you have strong skills with things. Finally, if you like to create organizational charts or develop long-range planning guides, you have strong skills with information or data.

Use Table 6-2 to Table 6-4 to help you decide what skills you already possess and how you enjoy using them. Each section—Skills with People, Skills with Things, and Skills with Information or Data—provides a list of verbs that describe how skills fit each context. For instance, under "Skills with Information or Data" are the verbs "planning", "researching", and "designing". Keep these verbs in mind so that you can use them to describe your own skills when it comes time to write your résumé.

Table 6-2 Skills with People

• Advising • Amusing • Communicating • Consulting • Counseling • Deciding • Feeling • Founding • Leading • Managing	• Negotiating • Performing • Persuading • Sensing • Serving • Taking Instructions • Treating • Working with Animals

Table 6-3 Skills with Things

• Emptying • Feeding • Handling (Objects) • Minding • Operating Equipment • Operating Vehicles	• Precision Working • Repairing • Setting Up • Using Tools • Working with the Earth or Nature

Table 6-4 Skills with Information or Data

• Achieving • Adapting • Analyzing • Comparing • Computing • Copying • Creating • Designing • Developing • Evaluating	• Expediting • Improving • Observing • Organizing • Planning • Researching • Retrieving • Storing • Synthesizing

6.3.3 Identifying Jobs

How do you know which jobs to apply for? Where do you look for jobs? Once you find a job, how can you get more information about that job?

Once you've completed your self-analysis and thought about your interests and skills, you will be much better prepared to know which jobs you should apply for. For instance, if you discover that you enjoy working with your hands and prefer to work outdoors, you would probably apply for a job as a landscaper or a carpenter but not for a job as a receptionist.

Now that you have a better idea what kind of jobs interest you and match your specific set of skills, you're ready to look for actual job openings. The following page on this section will get you started on where to look. Remember: there doesn't have to be a job opening in order for you to approach a company you'd like to work for. By sending your résumé to an employer and asking him or her to keep it on file, you potentially get your foot in the door. (Always call ahead and get the name of the employer so that you send your résumé to an actual person.) Even better, if you set up an appointment for an "information interview" with someone in the company, you can sit down face-to-face with someone and ask him or her questions about how you can make yourself a strong candidate for a potential opening. If you impress a specific individual in the company with your interest, enthusiasm, and abilities, you have a much better chance of being interviewed for a position when one becomes available. If you set up an information interview, try to meet with a person who has the kind of job you want or one who hires people in the position that you'd most like to have.

If you find a job description that you're interested in, there are a number of ways you can learn more about the position and precisely craft your résumé to fit the job.

- Make phone calls
- Talk to a current employee
- Ask questions

Don't feel as though you're limited to the information published in a job announcement. Remember: there's nothing wrong with calling the company to find out more about the job qualifications and the kind of person they're looking for. You can request to speak with a current employee, especially someone who has a similar position to the one they're advertising. And don't feel shy about asking pointed questions. Table 6-5 and Table 6-6 will give you hints on what sort of questions you might ask.

Table 6-5 Get to Know the Company

• Size
• Location
• Reputation
• Type of Product or Service
• Organizational Structure
• Training Program
• Unique Qualities/Programs
• Transfer and Promotion Policies

Table 6-6 Get to Know the Position

• What skills are needed?
• What are the specific duties?
• What on-the-job training is necessary/ available?
• Is creativity encouraged?
• Is cross-training available?
• What time is your shift?
• Is overtime available or expected?
• What tools or materials will be used?
• Will you work independently or with a team?
• Does the workload change throughout the year?

6.3.4 Writing Your Résumé

1. Work Experience Record and Résumé Categories

The next two sheets, Table 6-7 and Table 6-8 tell you precisely what information to include on your résumé draft. For each significant job that you've held, fill out a "Work Experience Record". Keep these records on file so that you'll always have this information and won't need to look it up again. Once you've filled out your records, your résumé should be organized around the following categories.

- Personal data
- Objective

- Education
- Experience
- Other categories (depending on the job description) such as qualifications or technical skills, language ability, military experience, or related activities
- References

Table 6-7 Work Experience Record

Name of Business:	
Address:	
Telephone:	
Dates of Employment: From	To
Job Title:	
Name of Supervisor:	
Rate of Pay: Beginning	Final
Reason for Leaving:	
Duties:	
Abilities Acquired:	
Best Part of the Job:	
Worst Part of the Job:	

Table 6-8 Résumé Categories

Personal Data: Full name, Address, Telephone Number, Cell Phone	Objective: Focus of résumé and employment goal. What type of employment are you seeking

Continued Table 6-8

Education: What, where, when? Do you have technical training? Any program certificates? Always present most recent first	Experience: Include all relevant employment, volunteer work, and internships. List the skills and responsibilities, dates, location (city/state), company
Other Categories: Qualifications or Technical Skills, Language Ability (read, write, speak), Military Experience, Related Activities	References: List 3-5 people. Should be educational, community based, or business. Be prepared to provide the person's name, mailing address, telephone number, and position

2. Action Words

Most résumés include brief descriptions of your work skills from prior jobs, and it's important to use action words when you write these descriptions. For instance, it's better to say that you taught children as a daycare worker than to say that you were with children all day. The verb "taught" shows you interacting with your work environment and describes a specific skill that you possess: in this case, the ability to teach others. Use Table 6-9 to help you decide what words best describe the work you've done in the past.

Table 6-9　Action Words for Strong Résumés

Achieved	Consulted	Expressed	Mentored	Reduced
Acquired	Contracted	Facilitated	Met	Referred
Acted	Contributed	Financed	Modeled	Related
Adapted	Converted	Fixed	Modified	Reported
Addressed	Cooperated	Followed	Monitored	Researched
Administered	Coordinated	Formulated	Negotiated	Responded
Advertised	Copied	Gained	Observed	Restored
Advised	Correlated	Gathered	Obtained	Reviewed
Advocated	Counseled	Gave	Offered	Saw
Aided	Created	Guided	Operated	Scanned

Continued Table 6-9

Analyzed	Dealt	Handled	Ordered	Scheduled
Answered	Decided	Headed	Organized	Screened
Anticipated	Defined	Helped	Overcame	Set Goals
Applied	Delegated	Identified	Oversaw	Shaped
Approved	Delivered	Implemented	Participated	Solved
Assembled	Designed	Improved	Perfected	Spoke
Assessed	Determined	Inclined	Performed	Strategized
Assisted	Developed	Indicated	Persuaded	Streamlined
Audited	Diagnosed	Influenced	Planned	Strengthened
Began	Directed	Inspected	Predicted	Stressed
Brought	Discovered	Interpreted	Prepared	Studied
Budgeted	Displayed	Interviewed	Presented	Succeeded
Built	Documented	Introduced	Prioritized	Summarized
Calculated	Drafted	Invented	Produced	Synthesized
Cared	Drove	Inventoried	Programmed	Supervised
Charged	Edited	Investigated	Projected	Supported
Checked	Eliminated	Judged	Promoted	Surveyed
Clarified	Enforced	Kept	Protected	Sustained
Classified	Enlisted	Learned	Proved	Talked
Coached	Ensured	Led	Provided	Taught
Collaborated	Established	Lifted	Publicized	Told
Collected	Estimated	Listened	Published	Trained
Comforted	Evaluated	Located	Purchased	Translated
Communicate	Examined	Logged	Questioned	Upgraded
Compared	Exceeded	Made	Raised	Utilized
Completed	Excelled	Maintained	Ran	Validated
Composed	Expanded	Managed	Ranked	Verified
Computed	Experimented	Mapped	Read	Visualized
Conceived	Explained	Mastered	Recorded	Won
Conducted	Explored	Mediated	Received	Wrote

3. Adaptive Skill Words

Résumés often list words that describe your personal traits, especially words that show what kind of adaptive skills you possess. Adaptive skills mean personal traits that allow you to interact with your work environment in a positive, effective manner. For example, you might want to describe yourself as an "efficient", "reliable", and "well-organized" worker rather than simply as an employee who "works hard". The more specific your adaptive skills are the better picture a potential employer has of you and your abilities. Use Table 6-10 to help you choose words that best describe your personal workplace traits.

Table 6-10 Adaptive Skill Words to Describe Your Personal Traits

Active	Effective	Outgoing
Adaptable	Efficient	Personable
Adept	Energetic	Pleasant
Aggressive	Enterprising	Positive
Analytical	Enthusiastic	Practical
Assertive	Exceptional	Productive
Committed	Experienced	Reliable
Competent	Fair	Resourceful
Conscientious	Firm	Self-confident
Cooperative	Forceful	Self-motivated
Creative	Hardworking	Self-reliant
Dedicated	Honest	Sensitive
Dependable	Independent	Sharp
Determined	Innovative	Sincere
Diligent	Logical	Strong
Diplomatic	Loyal	Successful
Disciplined	Mature	Tactful
Discreet	Objective	Tenacious
	Open minded	Well-organized

4. Revising and Proofreading

Once you've finished a draft of your résumé, it's important to make sure that the document looks good, makes sense, and contains no errors. An applicant with the most impressive qualifications in the world will not be considered as a potential employee if his or her résumé is vague or hard to read or contains surface errors.

Table 6-11 will help you look for aspects to revise or correct on your résumé draft, including the following aspects.

- Generic skill descriptions
- Self-deprecatory or wishy-washy self presentation
- Irrelevant information
- Dull or boring skill descriptions
- Fonts or styles that are too busy and hard to read
- A résumé that is long-winded and takes up too much space
- A résumé that is gaudy, using brightly colored paper, colored ink, or silly graphics
- Errors in spelling, grammar, and punctuation

Once you've written a clean, clear, and complete version of your résumé, show it to lots of people and get their feedback: ask friends, colleagues, family members, and teachers. You need to make sure that people from many different work environments and perspectives understand your résumé and think it presents you and your skills in the best possible light.

Finally, after you send out your résumé and get a job interview, ask a friend, colleague, family member, or teacher to do a practice interview with you. Make this interview as "real" as possible: dress up, hold it in an office (or make up part of your home to resemble an office, such as clearing the kitchen table and putting a notepad and a pencil on it), and don't let your friend, colleague, family member, or teacher "let down" his or her disguise as your potential "boss". Give them a copy of your résumé before the practice interview, and tell them to ask you questions about the information contained in it. After the practice interview is over, ask them to give you feedback on what you did well and what you need to improve. Remember: your résumé is not the only important part of getting a job. In an interview, you need to be able to talk clearly and distinctly about your past work skills and your specific abilities. You also need to avoid distracting gestures or body movements like bouncing or swinging your leg, pulling on your hair, or twiddling your thumbs. Finally, you should try to avoid using speech tags such as "like", "you know", or "uh", although these are hard habits to break. they make you sound young or unsure.

Once you've written your résumé, researched companies and job descriptions, tailored your résumé to fit specific job announcements, gotten a job interview, and practiced your interview with a friend, colleague, family member, or teacher. Then you're ready for the real thing. Good luck!

Table 6-11 Watch Out for these Mistakes

Too Snazzy!	Use only quality white or résumé paper Black ink only Limit graphics and formats
Forgot to Proofread!	Correct all spelling, grammar, and typographical errors
Irrelevant Info!	Customize your résumé according to the type of job
Overly Generic!	Add Specifics Include job duties and skills
Too Long!	One page is best; don't go beyond two pages
Boring!	Use action verbs and adjectives
Too Modest!	Give the best, most complete presentation of yourself. Avoid falsification and misrepresentation
Hard to Read!	Use a standard size 12 font. Don't use too many type styles

6.4 How to Write Business Letters

Typically, a business letter is reserved for only the most important of job-related or other professional communications: recommendation letters, cover letters, resignation letters, legal correspondence, company communications, etc. Since it's such a formal mode of communication, you'll want to make sure you have all of the formatting in place correctly. That's especially true if you're sending a hard copy to the recipient rather than an email.

The following sample letter format includes the information you need to include when writing a letter, along with advice on the appropriate font, salutation, spacing, closing, and signature for business correspondence.

1. Sample Letter Format

Contact Information (*Your contact information. If you are writing on letterhead that includes your contact information, you do not need to include it at* *the start of the letter.)*

Your name
Your address
Your city, state, zip code
Your phone number
Your email address
Date

Contact Information (*The person or company you are writing to.*)

Name

Title

Company

Address

City, state, zip code

Greeting

Dear Mr. /Ms. Last Name: (*Use a formal salutation, not a first name, unless you know the person extremely well. If you do not know the person's gender, you can write out his or her full name. For instance, you could write "Dear Pat Crody" instead of "Dear Mr. Crody" or "Dear Ms. Crody". Note that the person's name is always followed by a colon (:) in a business letter, and not a comma. If you do not know the recipient's name, it is still common (and safe) to use the old-fashioned "To Whom It May Concern:"*).

Body of Letter

When writing a letter, your letter should be simple and focused, so that the purpose of your letter is clear.

Single space your letter and leave a space between each paragraph. Left justify your letter. Use a plain font like Arial, Times New Roman, Courier New, or Verdana. The font size should be 10 or 12 points.

Business letters should always be written on white bond paper rather than on colored paper or personal stationary.

The first paragraph of your letter should provide an introduction as to why you are writing so that your purpose is obvious from the very beginning.

Then, in the following paragraphs, provide more information and specific details about your request or the information you are providing.

The last paragraph of your letter should reiterate the reason you are writing and thank the reader for reviewing your request. If appropriate, it should also politely ask for a written response or for the opportunity to arrange a meeting to further discuss your request.

Leave a blank line after the salutation, between each paragraph, and before the closing.

Closing

Best Regards,

Signature

Handwritten signature (*for a hard copy letter — use blue or black ink to sign the letter*)

Typed signature

2. Tips for Writing a Business Letter

Once you have written your business letter, proofread it (using spell check) on the screen. Then print it out and read it through at least one more time, checking for any errors or typos. (It's often easier to spot errors on a hard copy.)

Be on the lookout for formatting errors as well, such as two paragraphs that do not have a space in between, or lines that are indented incorrectly. Before putting your letter in an envelope, don't forget to sign above your typed name, using blue or black ink.

If you are using Microsoft Word or another word processing program to write your letter, there are templates available that can help you format your letter correctly.

EXERCISE 13: Business Letters Translation

Please read the following letters and translate the English ones into Chinese, or the Chinese one into English.

A letter of thanks

Dear Sirs:

We acknowledge with thanks receipt of your letter dated March 22, and the enclosed order. Their full contents have obtained our immediate attention. We have carefully noted all the specification shown on your order and happy to see that our cooperation has been further promoted.

We are enclosing herewith our sales confirmation in duplicate. Please kindly sign and return a copy with your duly signature. In addition, we would like to remind you to open a letter or credit in our favor in time so that we can arrange production and shipment as soon as possible.

We will do our best to maintain the good quality or our products, and will make the best effort to meet your special requirements. We sincerely hope that you will place further orders with us very soon if the execution of this order proves satisfactory to you.

Sincerely yours,
ABC Company

A letter of reply

尊敬的先生：

贵方 3 月 24 日来函及销售确认书附件收悉。我方对你方的立刻关照表示感谢。

按照你方的要求，我方将签署好的确认书复本寄送给你方。我方已经通过中国银行开具了以你方为受益人的信用证，该信用证完全符合所规定的条件及出货要求。两份文件均随信附上，请查收。

我方希望这次合作能够进一步深化我们之间的互惠伙伴关系。

DEL 公司谨上

A letter of reply

Dear Sirs:

We have received your letter dated on March 24, and the enclosed sales confirmation. Your immediate attention is greatly appreciated.

We are sending to you the copy of sales confirmation with our signature and the L/C in your favor. We hope that you execute this order with the best attention and promptness. Furthermore, when the goods are ready for shipment, please send us shipping advice by fax to facilitate us to make insurance arrangement. You can rest assured that we will place larger orders if your execution of this one proves to be satisfactory.

Sincerely yours,
DEL company

References

[1] University of Limerick. Bachelor/Master of Engineering in Mechanical Engineering [EB/OL]. [2017-12-10]. http://www3. ul. ie/courses/MechanicalEngineering. php.

[2] Mechanical Engineering at the University of Michigan. Admissions[EB/OL]. [2017-12-10]. https://me. engin. umich. edu/academics/undergrad/admissions.

[3] Massachusetts Institute of Technology. Department of Mechanical Engineering [EB/OL]. [2017-12-10]. http://catalog. mit. edu/schools/engineering/mechanical-engineering/#mechanical-engineering-bs-course-2.

[4] Princeton University. Programs [EB/OL]. [2017-12-10]. http://mae. princeton. edu/undergraduate/programs.

[5] Roser Christoph. "Faster, Better, Cheaper" in the History of Manufacturing: From the Stone Age to Lean Manufacturing and Beyond[J]. Florida: CRC Press, 2016.

[6] 大卫 G. 乌尔曼. 机械设计过程[M]. 北京:机械工业出版社,2010.

[7] Fradig G E, West G B. Assist Students in Developing Technical Reading Skills [M]. Goergia: American Association for Vocational Instructional Materials, 1985.

[8] Kalpakjian S, Schmid S R. Manufacturing Engineering and Technology-Hot Processes[M]. 北京:机械工业出版社,2004.

[9] Sclater N. Mechanisms and Mechanical Devices Sourcebook[M]. New York: McGraw-Hill, 2001.

[10] Skakoon J G. The Elements of Mechanical Design[M]. New York: ASME Press, 2008.

[11] Hans K B. CNC Handbook[M]. New York: McGraw-Hill, 2012.

[12] Sheehan M J. Word Parts Dictionary[M]. Carolina: MaFarland & Company, Inc, Publishers, 2000.

[13] Danaan Metge. Modeling and Simulating Mechanical Systems on a Transforming Dicycle[EB/OL]. [2017-12-08]. https://www. mathworks. com/company/newsletters/articles/modeling-and-simulating-mechanical-systems-on-a-transforming-dicycle. html.

[14] Wickert J. An Introduction to Mechanical Engineering[M]. 2nd. Cengage Learning,2006.

[15] Koenig D T. Manufacturing Engineering:Principles for Optimization[M]. 3rd. New York:ASME Press,2007.

[16] Trend E M, Wright P K. Metal Cutting[M]. 3rd. Boston: Butterworth-Heinemann,2000.

[17] Serope K,Steven R S,Hamidon M. 制造工程与技术:机加工[M]. 王先逵,改编. 北京:机械工业出版社,2012.

[18] Tracy Staedte. Advanced Vision Algorithm Helps Robots Learn to See in 3D [EB/OL]. [2017-12-10]. https://www. livescience. com/59878-vision-algorithm-helps-robots-see-in-3d. html.

[19] Billingsley J. Essentials of mechatronics [M]. New Jersey: Wiley-Interscience,2006.

[20] Srout K. Quality control in automation [M]. Boston, MA: Springer US,1985.

[21] UEfAP. Mathematical and scientific symbols [EB/OL]. [2017-12-10]. http://www. uefap. com/speaking/symbols/symbols. htm.

[22] Stephanie. How to Read Symbols and Equations in Calculus[EB/OL]. [2017-10-22]. http://www. statisticshowto. com/how-to-read-symbols-and-equations-in-calculus/.

[23] 朱派龙. 图解机械制造专业英语[M]. 1 版. 北京:化学工业出版社,2009.

[24] LanceMakes. Mechanical Hand [EB/OL]. [2014-1-16]. http://www. instructables. com/id/Mechanical-Hand-4/.

[25] Chan Ken,许忠能. 科技英语:中国学生专业英语应用指南[M]. 北京:清华大学出版社,2009.

[26] Heather Frazier,Jennifer Cognard-Black. Getting the Job: How To Write Your Resume. A Resume Writing Workshop[M]. Columbus:Ohio State University Center for the Study and Teaching of Writing,1999.

[27] Alison Doyle. Format for Writing a Business Letter[EB/OL]. [2018-2-8]. https://www. thebalance. com/sample-letter-format-2063479.

[28] 朱子熹. 世界 500 强员工必用的英文 E-mail 大全[M]. 北京:中国水利水电出版社,2011.